FRONTIERS OF SCIENCE

CHEMISTRY

Notable Research and Discoveries

KYLE KIRKLAND, PH.D.

Facts On File
An imprint of Infobase Publishing

CHEMISTRY: Notable Research and Discoveries

Facts On File, Inc.
An imprint of Infobase Publishing
132 West 31st Street
New York NY 10001

Library of Congress Cataloging-in-Publication Data

Kirkland, Kyle.
    Chemistry : notable research and discoveries / Kyle Kirkland.
        p. cm.—(Frontiers of science)
    Includes bibliographical references and index.
    ISBN 978-0-8160-7440-2
    1. Chemistry—Research. 2. Discoveries in science. I. Title.
    QD40.K485 2010
    540—dc22   2009024282

Facts On File books are available at special discounts when purchased in bulk quantities for businesses, associations, institutions, or sales promotions. Please call our Special Sales Department in New York at (212) 967-8800 or (800) 322-8755.

You can find Facts On File on the World Wide Web at http://www.factsonfile.com

Text design by Kerry Casey
Illustrations by Dale Williams
Photo research by Tobi Zausner, Ph.D.
Composition by Mary Susan Ryan-Flynn
Cover printed by Bang Printing, Inc., Brainerd, Minn.
Book printed and bound by Bang Printing, Inc., Brainerd, Minn.
Date printed: May 2010
Printed in the United States of America

10 9 8 7 6 5 4 3 2 1

CONTENTS

PREFACE

Discovering what lies behind a hill or beyond a neighborhood can be as simple as taking a short walk. But curiosity and the urge to make new discoveries usually require people to undertake journeys much more adventuresome than a short walk, and scientists often study realms far removed from everyday observation—sometimes even beyond the present means of travel or vision. Polish astronomer Nicolaus Copernicus's (1473–1543) heliocentric (Sun-centered) model of the solar system, published in 1543, ushered in the modern age of astronomy more than 400 years before the first rocket escaped Earth's gravity. Scientists today probe the tiny domain of atoms, pilot submersibles into marine trenches far beneath the waves, and analyze processes occurring deep within stars.

Many of the newest areas of scientific research involve objects or places that are not easily accessible, if at all. These objects may be trillions of miles away, such as the newly discovered planetary systems, or they may be as close as inside a person's head; the brain, a delicate organ encased and protected by the skull, has frustrated many of the best efforts of biologists until recently. The subject of interest may not be at a vast distance or concealed by a protective covering, but instead it may be removed in terms of time. For example, people need to learn about the evolution of Earth's weather and climate in order to understand the changes taking place today, yet no one can revisit the past.

Frontiers of Science is an eight-volume set that explores topics at the forefront of research in the following sciences:

- biological sciences
- chemistry

- computer science
- Earth science
- marine science
- physics
- space and astronomy
- weather and climate

The set focuses on the methods and imagination of people who are pushing the boundaries of science by investigating subjects that are not readily observable or are otherwise cloaked in mystery. Each volume includes six topics, one per chapter, and each chapter has the same format and structure. The chapter provides a chronology of the topic and establishes its scientific and social relevance, discusses the critical questions and the research techniques designed to answer these questions, describes what scientists have learned and may learn in the future, highlights the technological applications of this knowledge, and makes recommendations for further reading. The topics cover a broad spectrum of the science, from issues that are making headlines to ones that are not as yet well known. Each chapter can be read independently; some overlap among chapters of the same volume is unavoidable, so a small amount of repetition is necessary for each chapter to stand alone. But the repetition is minimal, and cross-references are used as appropriate.

Scientific inquiry demands a number of skills. The National Committee on Science Education Standards and Assessment and the National Research Council, in addition to other organizations such as the National Science Teachers Association, have stressed the training and development of these skills. Science students must learn how to raise important questions, design the tools or experiments necessary to answer these questions, apply models in explaining the results and revise the model as needed, be alert to alternative explanations, and construct and analyze arguments for and against competing models.

Progress in science often involves deciding which competing theory, model, or viewpoint provides the best explanation. For example, a major issue in biology for many decades was determining if the brain functions as a whole (the holistic model) or if parts of the brain carry out specialized functions (functional localization). Recent developments in brain imaging resolved part of this issue in favor of functional localization by showing that specific regions of the brain are more active during certain tasks. At the same time, however, these experiments have raised other questions that future research must answer.

The logic and precision of science are elegant, but applying scientific skills can be daunting at first. The goals of the Frontiers of Science set are to explain how scientists tackle difficult research issues and to describe recent advances made in these fields. Understanding the science behind the advances is critical because sometimes new knowledge and theories seem unbelievable until the underlying methods become clear. Consider the following examples. Some scientists have claimed that the last few years are the warmest in the past 500 or even 1,000 years, but reliable temperature records date only from about 1850. Geologists talk of volcano hot spots and plumes of abnormally hot rock rising through deep channels, although no one has drilled more than a few miles below the surface. Teams of neuroscientists—scientists who study the brain—display images of the activity of the brain as a person dreams, yet the subject's skull has not been breached. Scientists often debate the validity of new experiments and theories, and a proper evaluation requires an understanding of the reasoning and technology that support or refute the arguments.

Curiosity about how scientists came to know what they do—and why they are convinced that their beliefs are true—has always motivated me to study not just the facts and theories but also the reasons why these are true (or at least believed). I could never accept unsupported statements or confine my attention to one scientific discipline. When I was young, I learned many things from my father, a physicist who specialized in engineering mechanics, and my mother, a mathematician and computer systems analyst. And from an archaeologist who lived down the street, I learned one of the reasons why people believe Earth has evolved and changed—he took me to a field where we found marine fossils such as shark's teeth, which backed his claim that this area had once been under water! After studying electronics while I was in the air force, I attended college, switching my major a number of times until becoming captivated with a subject that was itself a melding of two disciplines—biological psychology. I went on to earn a doctorate in neuroscience, studying under physicists, computer scientists, chemists, anatomists, geneticists, physiologists, and mathematicians. My broad interests and background have served me well as a science writer, giving me the confidence, or perhaps I should say chutzpah, to write a set of books on such a vast array of topics.

Seekers of knowledge satisfy their curiosity about how the world and its organisms work, but the applications of science are not limited to intellectual achievement. The topics in Frontiers of Science affect society on a multitude of levels. Civilization has always faced an uphill battle to procure scarce resources, solve technical problems, and maintain order. In modern times, one of the most important resources is energy, and the physics of fusion potentially offers a nearly boundless supply. Technology makes life easier and solves many of today's problems, and nanotechnology may extend the range of devices into extremely small sizes. Protecting one's personal information in transactions conducted via the Internet is a crucial application of computer science.

But the scope of science today is so vast that no set of eight volumes can hope to cover all of the frontiers. The chapters in Frontiers of Science span a broad range of each science but could not possibly be exhaustive. Selectivity was painful (and editorially enforced) but necessary, and in my opinion, the choices are diverse and reflect current trends. The same is true for the subjects within each chapter—a lot of fascinating research did not get mentioned, not because it is unimportant, but because there was no room to do it justice.

Extending the limits of knowledge relies on basic science skills as well as ingenuity in asking and answering the right questions. The 48 topics discussed in these books are not straightforward laboratory exercises but complex, gritty research problems at the frontiers of science. Exploring uncharted territory presents exceptional challenges but also offers equally impressive rewards, whether the motivation is to solve a practical problem or to gain a better understanding of human nature. If this set encourages some of its readers to plunge into a scientific frontier and conquer a few of its unknowns, the books will be worth all the effort required to produce them.

ACKNOWLEDGMENTS

Thanks go to Frank K. Darmstadt, executive editor at Facts On File, and the staff for all their hard work, which I admit I sometimes made a little bit harder. Thanks also to Tobi Zausner for researching and locating so many great photographs. I also appreciate the time and effort of a large number of researchers who were kind enough to pass along a research paper or help me track down some information.

INTRODUCTION

Early scientists and philosophers invested much effort in the search for the fundamental substance or substances—the simplest kind of matter that comprises the world and all of its various materials. The ancient Greek philosopher Thales (ca. 635–556 B.C.E.) postulated that water is the fundamental substance. Although this idea may not sound realistic today, the hypothesis was a reasonable one. Life depends on water and Earth contains a huge quantity of water in oceans and rivers; water falls from the sky as rain and seeps through the ground in wells. The ancient Greek philosopher Empedocles (ca. 495–435 B.C.E.) expanded the list of fundamental substances to four—water, earth, air, and fire.

Key discoveries beginning in the late 18th century sparked a major advance. The frontiers of chemical knowledge at that time involved the nature of the bewildering array of substances and materials in the world. In the 1770s and 1780s, the French chemist Antoine-Laurent Lavoisier (1743–94) developed techniques to measure and quantify substances and their components with a great deal of precision. A few decades later, the British researcher John Dalton (1766–1844) formulated his atomic theory, which explained why *elements* combined with one another in fixed ratios to make *compounds.* Then, in the late 1860s, the Russian chemist Dmitry Mendeleyev (1834–1907) noted the similarities in the chemical properties of certain groups of elements and drew up an early version of the periodic table of chemical elements. A modern version of the table can be found in this book's appendix. These discoveries brought attention to fundamental units of matter—*atoms* and *molecules.*

Elements combine in various ways, leading to a large number of compounds that exhibit a remarkable variety of properties. Today, pioneering researchers are exploring these properties, and learning how to make materials that are smaller, stronger, and more adaptable than ever before. *Chemistry*, one volume in the multivolume Frontiers of Science set, is about explorers and scientists who venture into the unknown frontiers of chemistry—and quite often find materials exhibiting remarkable or useful properties. Some of these materials consist of large molecules that can cure diseases, while others comprise motors or machines so small that they are made of only a few atoms.

Each chapter of this book explores one of these frontiers. Reports published in journals, presented at conferences, and described in press releases illustrate the kind of research problems of interest in chemistry and how scientists attempt to solve them. This book summarizes a selection of these reports—unfortunately there is room for only a fraction of them—that offers students and other readers insight into the methods and applications of chemistry.

Students need to keep up with the latest developments in these quickly moving fields of research, but they have difficulty finding a source that explains the basic concepts while discussing the background and context essential for the "big picture." This book describes the evolution of each of the six main topics it covers, and it explains the problems that researchers are currently investigating as well as the methods they are developing to solve them.

Chapter 1 discusses the challenges and opportunities of making new substances. The traditional means of acquiring novel materials is by trial and error, where the experimenter tries one substance, and if it fails to meet the required specifications, tries another. Faster methods would employ the principles of how atoms combine and the properties that different combinations bestow upon the product. Although these are extremely difficult methods to develop, a better understanding of the science of materials would lead to the ability to design substances as needed, instead of relying on chance discoveries. It might also open up new engineering possibilities, such as an elevator tall enough to lift satellites into space.

Chapter 2 zooms in on the possible development and use of tiny engines. Objects in the familiar world obey Newton's laws, at least approximately, but on the atomic or molecular scale, Newton's laws do not apply, and advanced concepts such as *quantum mechanics* take hold. Gravity is a prominent force in most people's experience—any slip and fall serves as a painful reminder—but in the tiny world of nano-technology, forces other than gravity predominate. These unusual facets of nanotechnology pose problems but also create unique situations that researchers are beginning to exploit.

Other frontiers of chemistry take researchers into hidden realms. One of these areas is the human skull, which houses three pounds (1.4 kg) of one of the most complex substances in the universe—the human brain. Chemical investigations of the brain have taken even longer to develop than other fields of chemistry, due to the relative inaccessibility of brain tissue. Yet researchers are making strides in this difficult field, including the development of medications to alleviate disorders such as depression and drug abuse, as explained in chapter 3.

In addition to designing new substances and expanding technology to miniature scales, chemists and engineers are looking to create materials that are responsive to the environment. Most materials, such as the wings of an airplane, have fixed properties, yet must perform under varying conditions of pressure and force. This means the object, such as a wing, must be designed to fulfill a variety of roles, resulting in compromises and inefficiencies because the properties are usually not optimal for any of the conditions. Materials that are "smart," in the sense that they adapt to their environment, would alleviate this problem, and create opportunities to build extremely useful and "intelligent" systems. Chapter 4 discusses research efforts in this field.

Encountering and solving problems are a routine part of chemistry. Chapter 5 discusses the serious problem of energy resources, and a possible solution. Oil, coal, and natural gas account for much of the world's energy, yet these sources are rapidly being depleted, and the by-products of their use, such as pollution and carbon dioxide emissions, are damaging the environment. Devices known as fuel cells, which generate electricity by specific chemical reactions, offer a promising alternative—a clean and nearly inexhaustible source of energy—if researchers can overcome certain obstacles.

Other interesting and important problems revolve around the passage of time, which obscures the history of past human societies, cloaking their successes and failures in mystery. Archaeologists and historians who study fossils and written records have learned much about how people used to live and work. But chemistry can expand this frontier of knowledge by revealing more about the nature of the objects that have been left behind. Chapter 6 focuses on this topic. For example, researchers have studied the clothing, tools, and body of a man who had been buried in ice for about 5,300 years and have determined his diet, where he spent his childhood, and what may have happened in the last few days of his life. Since many of the problems facing society today are similar to those that previous civilizations have encountered, learning from past successes, and mistakes, have potentially enormous benefits.

Few researchers have enjoyed the phenomenal successes of Lavoisier, Dalton, and Mendeleyev, yet the contributions, large and small, of many scientists add up to significant progress. The frontiers of chemistry are as exciting today as they were in the 18th and 19th centuries.

Developing and Designing New Chemicals and Materials

In the fairy tale *Jack and the Beanstalk,* magical beans produce a stalk that grows tall enough for a young man to climb all the way into the sky. Jack encounters a giant and, after some trickery, ends up with the giant's gold—at least in some versions of the tale. The story is clever and interesting, particularly the part about a beanstalk extending into space.

This old idea of a structure that attains phenomenal height has gained much interest lately. As a way of reaching space, a tower would be safer and more efficient than rocket launches. But there is a perplexing problem—finding a material that can stand up to great heights. Even the strongest steel would fail because a steel tower that high could not support its own weight; steel is strong but heavy.

Height has always been difficult to attain because strong and sturdy materials, such as stone, are heavy, and there is a limit to how much weight they can bear before they collapse. At its original height of about 481 feet (147 m), Cheops (Khufu) Pyramid (the Great Pyramid of Giza) was the tallest building for nearly 4,000 years. (The pyramid's present height is 450 feet [137 m] due to erosion.) The tallest completed human-built structure as of May 2009 is the CN Tower in Toronto, which is 1,815 feet (553 m) tall, although a skyscraper under construction in Dubai called the Burj Dubai has already surpassed the CN Tower and will be about 2,684 feet (818 m) when completed. But a tower that can transport satellites into space requires a height many times this figure. Attaining this height is impossible until researchers develop a material that is strong enough to support the structure, yet light enough so that the weight due to gravity does not cause it to collapse.

The same difficulty occurs when people design and build any object—a substance or chemical having the required properties must be available. Jet engines, for example, were only a gleam in the eye of aeronautics engineers before the development of high-strength, heat-resistant *alloys*. Medications to fight bacterial infections must be able to kill the invading microorganisms without harming the patient's own cells.

A lack of proper materials hampers the progress of technology in all fields. Discovering and developing new substances is presently a slow process, based on luck and intensive work. But now scientific research is pushing into a new frontier—designing and manufacturing chemicals and materials, rather than relying on chance discoveries.

INTRODUCTION

The ability to design new materials will require a deep understanding of chemistry. A person who designs an object must be able to predict its properties in advance, otherwise the "design" would be no more productive than a guess. One of the greatest steps toward this goal in chemistry came in 1869, when the Russian chemist Dmitry Mendeleyev (1834–1907) devised an early version of the periodic table of chemical elements. A modern version of the table appears in the appendix on page 200 of this book.

Elements are fundamental substances that cannot be broken down into smaller chemical components. The smallest unit of an element is an atom, a term based on the Greek word *atomos,* meaning indivisible. But atoms are divisible—they consist of a *nucleus* containing positively charged particles called *protons* and electrically neutral particles called *neutrons,* surrounded by a swarm of electrically negative particles called *electrons.* In chemical *reactions,* atoms interact and combine to form a molecule of a compound. (Chemical reactions also occur when the atoms in molecules interact and combine to form even bigger compounds.) For example, two atoms of hydrogen (H) and one atom of oxygen (O) react to form a molecule of the compound known as water, written chemically as H_2O. An element's chemical properties—the chemical reactions it enters and the products formed from these reactions—depend on the properties of the atom.

Mendeleyev noticed that when he listed the chemical elements known at the time in order of atomic weight—the mass of the atom—elements with similar properties appeared periodically. For instance, lithium, sodium, potassium, rubidium, and cesium are all soft metals that tend to engage in the same chemical reactions. Toss a pellet of pure sodium into water and it will strongly react with water, producing hydrogen gas. The same is true for the other elements of this group. When elements are arranged in a table, columns form groups that have similar chemical properties.

Mendeleyev observed some gaps in this early table. Rather than dismissing this observation as unimportant, Mendeleyev hypothesized that the gaps represented as yet undiscovered elements. Using the groups of the periodic table, he could predict some of the properties for these unknown elements. For example, in 1871, Mendeleyev predicted the existence of an element below silicon, which he called ekasilicon, and some of its properties, such as mass, density, and the nature of the compound it would produce when combined with oxygen. The German chemist Clemens Winkler (1838–1904) found this element in 1886. Germanium, as it was called (for Germany), has many of the properties Mendeleyev predicted.

Later versions of the periodic table ordered the elements by *atomic number,* a concept that had not been formulated in Mendeleyev's time. The atomic number is the number of protons in the nucleus of an atom. This number specifies each of the elements—for instance, oxygen has an atomic number of eight, and all atoms of oxygen have eight protons (though oxygen atoms may vary in the number of neutrons in the nucleus). As of May 2009, researchers have identified a total of 117 elements, though many of the larger elements are highly unstable—chemists make them during special experiments in the laboratory, but the atoms decay quickly.

The periodic table of elements gives order to the bewildering variety of substances found on Earth. People have been working with various materials for a long time, but the development of new substances was a slow and fortuitous process. About 30,000 years ago, the first *ceramic* appeared,

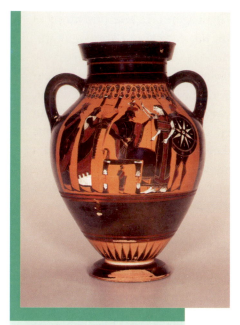

Sixth-century B.C.E. Greek ceramic vase *(Yale University Art Gallery/Art Resource)*

in the forms of pottery and small statues made by heating mineral-bearing substances such as clay. Some 10,000 years ago, people began working with metals, gradually developing the ability to use fire to extract copper from minerals such as malachite, and casting the molten metal into different shapes. Higher temperatures are required to extract and shape other metals such as iron, which was not extensively produced until about 3,500 years ago. In 1839, the American inventor Charles Goodyear (1800–60) developed a process known as vulcanization, which made rubber more durable and elastic, and a few years later, British engineer Sir Henry Bessemer (1813–98) discovered an efficient means of producing high-quality steel. These discoveries, among others, enabled manufacturers to make an enormous number of useful objects and aided the rise of industry—the Industrial Revolution—in the 19th century.

But lacking theoretical guidance, such discoveries were made by chance or by trial and error. New materials were not planned or designed; they were instead lucky finds or they were the result of painstaking efforts in which a researcher repeatedly concocted substances and tested their properties in a series of trials. Most new substances have serious flaws and fail in some way or another, forcing the researcher to try again. Goodyear and Bessemer needed years to perfect their processes. As Bessemer wrote in his autobiography, "Several weeks were sometimes necessary to make and fit up the apparatus required to test a new theory, and it too often happened that the first hour's trial of the new scheme dashed all the high expectations that had been formed, and we had again to retrace our steps. Thus, week after week went on amid a constant succession of newly-formed hopes and crushing defeats, varied with occasional evidences of improvement."

Bessemer was not alone in his troubles. Thousands of years had passed from the first metalworking to the development of reliable methods to produce steel.

The same slow progression has also been true for the development of medications. For instance, ancient Greeks and others knew of a powder extracted from the bark of willow trees that could help relieve fevers. In the early 19th century, the active ingredient—the component that produces the desired effect—had been found, and chemists used it to make salicylic acid. This chemical was effective in treating fever and pain but also generated side effects—undesirable effects such as bleeding. Decades later, chemists succeeded in developing another related chemical, acetylsalicylic acid, which proved to have superior properties. This substance went on sale in 1899 under the name aspirin.

Even the development of Mendeleyev's periodic table of chemical elements did not grant researchers the ability to plan and design new materials. Although Mendeleyev correctly predicted the existence of a few new elements, the number of compounds that can be made from elements is extremely vast and complex. Chemical Abstracts Service, a division of the American Chemical Society, maintains a registry of known substances. As of May 2009, there are about 47 million substances in this registry, and roughly 4,000 new substances are added every day.

The chemical properties of atoms and the manner in which they react to form compounds depend on the arrangements of their electrons. This complicated process involves advanced ideas in chemistry and physics, including quantum mechanics, which describes the behavior of small particles such as atoms. Some of these ideas are described in the following section. But even with this knowledge, along with the exciting possibility of quickly developing new and better materials, researchers are only beginning to understand the tools and techniques needed for material design. Molecules get many of their useful properties from their complex structures, held together with various kinds of chemical bonds.

CHEMICAL BONDS AND MOLECULAR STRUCTURE

Chemical bonds are attractive forces by which atoms join to form compounds. Explaining bonds with mathematical precision is the realm of quantum mechanics, but the basic idea involves the stability of certain arrangements of electrons. In a simplified description of an atom, electrons orbit around the nucleus, as illustrated in the figure. (Neutral atoms have the same number of electrons as protons.) These orbits are grouped in shells. As a general rule, many atoms are stable when they have eight electrons in their outer shell. This "octet rule" is true for gases such as neon, argon, and xenon, which rarely react with other elements. Other atoms engage in chemical reactions that tend to satisfy this octet rule; for example, oxygen, with six electrons in its outer shell, acquires at least a share of two more.

Strong bonds form when atoms transfer one or more electrons, in which case the bond is called an *ionic bond,* and when atoms share one or more electrons, in which case the bond is known as a *covalent bond.* The bond between each hydrogen atom and oxygen in H_2O is covalent—the oxygen atom shares an electron with each hydrogen atom. Oxygen has six electrons in its outer shell and hydrogen has one, so the result is that the oxygen atom obtains eight electrons in its outer shell (the usual six plus a share in two more) and the hydrogen atoms have two apiece. (Hydrogen is the smallest atom and is satisfied with only two electrons, as is helium, which also has two.) A simple example of an ionic bond occurs when sodium reacts with chlorine to form sodium chloride, also known as table salt. In this case, sodium donates the single electron in its outer shell to chlorine, which needs only one electron to fill its outer shell.

Another bond, similar to covalent bonds, occurs in metals, in which atoms share electrons with many of their neighbors. Other kinds of bonds are weaker than covalent, ionic, and metallic bonds, and often involve subtle interactions based on electric attraction or repulsion of the charged components of molecules. For instance, *proteins* are large molecules that perform a variety of functions in biological tissues and are formed by covalent bonding of a sequence of smaller molecules known as amino acids. The sequence of amino acids folds into a certain shape, which governs the protein's function. Keeping this shape intact is the job of many weak interactions among the hundreds or even thousands of atoms in the protein.

The structure formed by these bonded atoms is critical for many compounds. Diamond and graphite are two substances with vastly different properties—diamond is one of the hardest substances known, and graphite is soft and often used as a lubricant or as a pencil "lead" to leave black marks on paper. Yet both of these substances are composed of the same element! Diamond and graphite are different forms of carbon.

Why are the properties of diamond and graphite so different? Properties depend on the structure or configuration of the atoms. But to get a look at such structures, scientists have to resort to unusual tools. Because atoms are far too small to be seen with ordinary light-based microscopes, researchers use *X-rays,* a type of electromagnetic radiation that has a much smaller wavelength than light. Smaller wavelengths give X-rays better resolution—the ability to discern small details—although they also give X-rays higher energy, which makes them more dangerous than light.

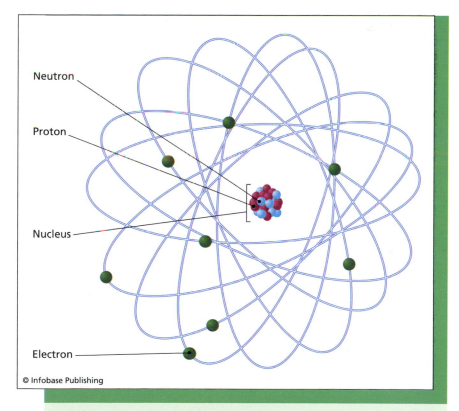

Neutron

Proton

Nucleus

Electron

© Infobase Publishing

In the simplest model of an atom, electrons swarm around a compact nucleus containing protons and neutrons.

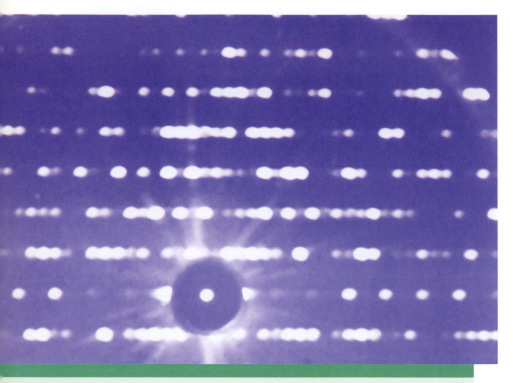

Certain mathematical techniques can determine molecular structure from X-ray diffraction images, such as this X-ray diffraction image of a protein. *(Alfred Pasieka/Photo Researchers, Inc.)*

    The first step in many techniques that visualize a substance's structure is to make a *crystal*. A crystal is a solid substance in which the molecules are packed in repeating, geometric arrangements. Many, though not all, substances will form crystals under certain conditions, and in general the process does not alter the structure of the molecules being crystallized. When bombarded with X-rays, a crystal scatters the radiation, but not in a random fashion. The direction and intensity of the scattered X-rays depends on the geometry of the crystal, and X-ray detectors record the resulting patterns. With a mathematical technique developed by the German physicist Max von Laue (1879–1960) and the British father and son team Sir William Henry Bragg (1862–1942) and Sir William Lawrence Bragg (1890–1971) in the 1910s, researchers can use the scattered X-ray pattern to deduce the structure. This technique is known as *X-ray crystallography*. Although the crystal consists of many atoms or molecules, they are all in the same orientation—the repeating geometry of the crystal—and so the configuration of a single unit can be determined. Ordinary solids that lack repeating geometry could not be used in this technique since the randomly oriented atoms would fail to create an interpretable pattern of scattered radiation.

Both graphite and diamond are made of carbon, but in graphite the bonded carbon atoms are stacked in a plane (a flat, two-dimensional figure), with only weak bonds between planes. Part B of the figure illustrates this structure. The carbon bonds within the plane are strong, but the planes slide past one another, giving graphite its soft, slippery property. A diamond, illustrated in Part A of the figure, has a much different configuration. In diamonds, the bonded carbon atoms do not form a plane but rather a three-dimensional figure known as a tetrahedron, and the packing is tight. As a result, diamonds are hard. Carbon squeezes into diamonds under high pressures and temperatures, such as in the depths of Earth, or in special laboratories designed to produce these extreme conditions. Un-

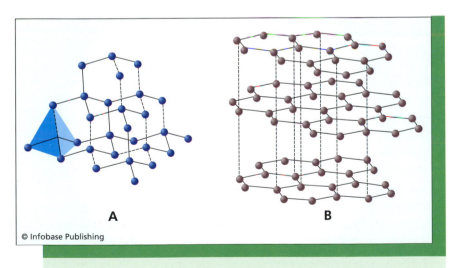

A   B

© Infobase Publishing

(A) In a diamond, carbon atoms form the shape of tetrahedrons, which are packed together. A single tetrahedron is shaded. Black dots represent carbon atoms. (B) Graphite consists of planes of carbon atoms.

der normal conditions at Earth's surface, graphite is the usual form—and diamonds will eventually become graphite when brought to the surface. But fortunately for people who pay a lot of money for these rare gems, the conversion process takes thousands of years to occur.

Diamonds are so hard that they cut many other materials, and they are used industrially for this purpose. But because diamonds are so expensive, people often search for cheaper substitutes. Knowledge of the simple structure that makes diamond so hard is useful in finding or making similar materials, and Hsiu-Ying Chung and Richard B. Kaner of the University of California, Los Angeles, and their colleagues recently published a paper describing an alternative to diamonds. The idea of Chung, Kaner, and their colleagues was to combine the dense though soft element rhenium with boron. In addition, the researchers synthesized (made) this

Lawrence Livermore National Laboratory

During World War II, the Los Alamos Laboratory in New Mexico conducted a highly secret and successful operation to build the world's first atomic bomb. The bomb ended the war, and politicians as well as scientists began to appreciate the benefits of establishing scientific laboratories and conducting research. The national laboratory at Los Alamos continued its work on atomic physics and other projects after the war, and in the early 1950s, the physicists Ernest Lawrence (1901–58) at the University of California, Berkeley, and Edward Teller (1908–2003), then at Los Alamos, urged the establishment of another laboratory. Under the supervision of the University of California, a new national laboratory was set up in Livermore, California, in 1952, with Herbert York (1921–2009), a former student of Lawrence, as its first director.

Early efforts focused on nuclear science and atomic technologies, similar to the laboratory at Los Alamos. In the substance, rhenium diboride, at standard atmospheric (ambient) pressure, so they did not have to apply extreme pressure, which requires costly procedures and equipment. As reported in "Synthesis of Ultra-Incompressible Superhard Rhenium Diboride at Ambient Pressure," published in *Science* in 2007, this combination formed short bonds and tight packing, achieving a hard, solid material. In the words of the researchers, "scratch marks left on a diamond surface confirmed its superhard nature"—the substance is hard enough to scratch diamond!

Scientists who are seeking new materials often think beforehand about what kind of structure would work best. But on occasion, a new and unexpected structure presents itself, as happened in 1985, when researchers found a third form of carbon that became known as buckminsterfullerene.

1950s, Livermore scientists developed a nuclear weapon that could be launched from submarines and weapons that carried multiple warheads. Livermore also investigated other aspects of nuclear weapons and the dangerous radiation released upon explosion. Nuclear radiation often damages an organism's genetic material, known as *deoxyribonucleic acid* (DNA), and researchers at Livermore and other laboratories began studying DNA in order to understand these effects. These efforts eventually led to the Human Genome Project, an ambitious undertaking begun formally in 1990 to map the whole set of human DNA. The Human Genome Project was finished in 2003, though much research on the data continues.

Today, the laboratory employs 3,500 scientists and has an annual budget of about $1.6 billion. In addition to projects such as Manaa's fullerene research, Lawrence Livermore National Laboratory operates the world's largest *laser* facility—National Ignition Facility—as well as a number of powerful supercomputers. This equipment provides researchers with the tools to study atoms and molecules on an experimental basis as well as supplying computational power to simulate their activity on a theoretical level.

BUCKMINSTERFULLERENE— A MOLECULE IN THE SHAPE OF A SOCCER BALL

In 1985, Harry Kroto, then a professor at the University of Sussex, in England, came to Rice University in Houston, Texas, to work with Richard Smalley and Robert Curl on a project involving the chemistry of carbon-containing molecules. The researchers used a mass spectrom-

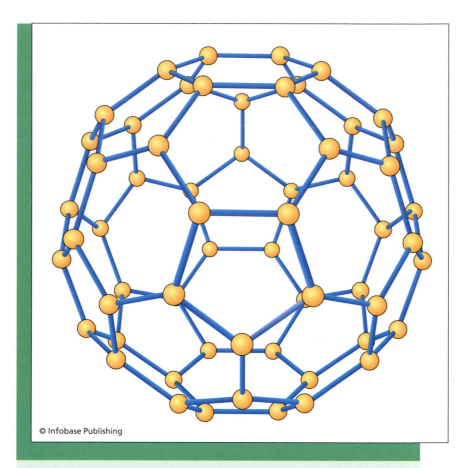

The structure of C_{60} consists of carbon atoms bonded together in the shape of a soccer ball. In the figure, the dots (or nodes) represent carbon atoms, and the lines represent bonds, some of which are double bonds in which two pairs of electrons are shared.

eter to identify the molecules; a mass spectrometer splits molecules into charged fragments, uses magnetic fields to measure the charge-to-mass ratio of the fragments, and sums up the masses to specify the original molecule. One of the main goals of the experiments was to study stellar chemistry, but then the researchers discovered to their surprise a molecule of carbon containing 60 atoms—C_{60}.

Such a large number of atoms could possibly adopt any one of a number of different arrangements. No one previously knew that pure carbon could assume another form other than graphite and diamond, but as other researchers carefully examined carbon sources such as soot (the remains of combustion), they found some C_{60}. In the early 1990s, curious scientists examined the structure of C_{60} using *nuclear magnetic resonance* (NMR) spectroscopy. Unlike X-ray crystallography, NMR spectroscopy does not require crystallization of the material. The process involves quantum mechanics, but in simple terms, it begins with a strong magnetic field to align the atoms of a sample of the material. Atoms respond to radio waves differently depending on their position in the molecule, and researchers map the molecule by transmitting and detecting these low-frequency electromagnetic waves. The technique is similar to magnetic resonance imaging (MRI), used by physicians to image a patient's body. NMR spectroscopy is often easier than X-ray crystallography but only works for certain molecules, and it generally produces a fuzzier picture.

NMR spectroscopy results suggested a soccer ball–shaped structure, as illustrated in the figure. Proof of this structure came in 1991, when Joel M. Hawkins, Frederick J. Hollander, and their colleagues at the University of California, Berkeley, crystallized C_{60} and studied it with X-ray crystallography. This technique, which generally offers exceptionally high resolution, confirmed the astonishing soccer ball frame.

C_{60} was named buckminsterfullerene, in honor of the visionary American architect Richard Buckminster Fuller (1895–1983). Fuller is known for developing and promoting the geodesic dome, which C_{60} resembles. (Buckminsterfullerene molecules are also sometimes called buckyballs.) Later, researchers discovered this molecule belongs to a family of related carbon structures, which have become known as fullerenes. The smallest fullerene is C_{20}, containing 20 carbon atoms.

The diameter of a C_{60} molecule is about 0.00000004 inches (0.0000001 cm). This distance is a *nanometer* (a billionth of a meter), and some scientists are working on miniature technology, known as nanotechnology, on this scale. (Nanotechnology is the subject of chapter 2 in this book.) A molecule having a structure that forms a cage, as does the soccer ball shape of buckminsterfullerene, is inviting to researchers who would love to insert small molecules such as medicines inside it and use C_{60} to transport and deliver these molecules to inaccessible places in the human body. These efforts have yet to be successful, but researchers are continuing to investigate the possibility.

Fullerenes are generally stable molecules, but they do enter into chemical reactions. Researchers have managed to replace some of the carbon atoms in buckminsterfullerene with other atoms such as nitrogen, making molecules such as $C_{48}N_{12}$. This molecule has a soccer ball shape but with 12 nitrogen atoms substituting for carbon. The chemistry of $C_{48}N_{12}$ is different from an all-carbon structure because nitrogen is more reactive than carbon—nitrogen is a primary component of many explosive compounds such as trinitrotoluene (TNT) and ammonium nitrate.

Researchers such as Riad Manaa at the Lawrence Livermore National Laboratory in California apply the principles of chemistry and physics to examine the properties of $C_{48}N_{12}$ as well as molecules that have not yet been made—but possibly could be, one day. Manaa and his colleagues are exploring $C_{48}B_{12}$, in which boron (B) replaces 12 of the carbons. $C_{48}N_{12}$ and $C_{48}B_{12}$ may interact in a way such that the molecules together could form fast electronic switches, similar to those used in computers, except on a molecular scale. The Lawrence Livermore National Laboratory is well equipped for studies on such tiny scales, as described in the sidebar on pages 10 and 11.

Studying new materials occupies much of the time of materials scientists. These materials can include substances that have not yet been found or made but are theoretically possible, opening up exciting new avenues in research. Finding and developing these materials then becomes the next challenge.

FINDING THE RIGHT STUFF

A cook without a recipe book is forced to experiment—a dash of salt here, a few onions there, vary the temperature, and then taste the result. In most trials, the result will probably not be a very palatable dish, but once in a while the chef may hit upon the gourmet equivalent of a jackpot.

The same trial-and-error process occurs in materials science. Sometimes researchers need to devise a method to manufacture a substance that theory has suggested may exist, given the right conditions. In other cases, an engineer or manufacturer may need a substance with specific properties, such as hardness or the ability to dissolve in certain liquids, and no known substances meet all the requirements. Lacking a specific set of instructions on how to make new materials, scientists of the past have relied on a little bit of inspiration—their knowledge of chemical principles, for instance—and a lot of perspiration as they try and try again, as Bessemer described in his autobiography.

Plastics are one example. Scientists use the term *plastic* to refer to polymers, which are materials composed of a chain of bonded molecules. For instance, most credit cards are made from a plastic known as polyvinyl chloride (PVC), which consists of linked molecules of vinyl chloride. (Old "vinyl" record albums were often made from the same material.) The first plastic made entirely in the laboratory (as opposed to natural products) was Bakelite, concocted around 1908 by the Belgian-American chemist Leo Baekeland (1863–1944).

One of the primary motivations for developing plastic was the burgeoning electronics industry of the late 19th century. Electrical engineers needed a flexible material that was not an electrical conductor, with which they could protect and insulate electrical wires and circuits. Chemists had been experimenting with materials such as phenol and formaldehyde for decades before Baekeland hit upon the right combination of these substances, as well as the right temperature and pressure at which they would combine to form a useable plastic.

Since the development of Bakelite, many other types of plastic have appeared. Plastics have a huge number of applications today, including bags and containers of all shapes and varieties.

Materials called *composites* are also in widespread use. A composite consists of small fibers of one substance embedded in a matrix of another substance. The fibers reinforce the matrix, adding stiffness. One of the first composites was fiberglass, in which thin fibers of glass reinforce a plastic material. Fiberglass is strong yet light in weight.

Composites fill many needs, especially the need for tough, lightweight materials used in cars and airplanes. Vehicles of the past have

A 1953 Corvette *(Car Culture/Corbis)*

conventionally been made of metal, a strong material that forms the body of the craft and protects the passengers in case of collision. But metals are dense and have much weight per unit volume. This weight adds to the bulk of the vehicle, making it sluggish, as well as unnecessarily increasing the consumption of fuel. One of the first automobiles to have a composite body was the Corvette sports car. Introduced in 1953, the Corvette had a fiberglass body. Aircraft manufacturers have also been turning to composites for the same reason, using less aluminum and more composites. Military jets, which rely on speed and maneuverability, contain a lot of composites—in many cases, about a third of the craft—and about half of Boeing's new 787 airliners will be made from composite material.

Popular high-strength composite materials today are commonly made with fibers of graphite. This may sound strange, considering graphite is so soft. But recall that the structure of a material is critical in determining its properties. Graphite is soft and slippery because its sheets of carbon are only weakly bonded; in carbon composites, the carbon is twisted into threads, which are held together with strong bonds. When these thin carbon fibers are glued into a piece of plastic, the result is a relatively lightweight material of exceptional strength.

Structure is also essential in complex biological molecules. A lot of medicines used for psychiatric illnesses such as depression rely on their ability to interact with certain proteins in the brain. For instance, a class of antidepressants—medications that alleviate the symptoms of depression—act on proteins involved with the collection (reuptake) of the chemical serotonin, and they are known as selective serotonin reuptake inhibitors (SSRIs). This class of antidepressants includes Prozac and Zoloft. Earlier medications were also effective and are still sometimes used though they produce a number of side effects, such as dietary problems. Although an SSRI can also generate potentially dangerous side effects, psychiatrists tend to observe these effects less often. (Brain chemistry is the subject of chapter 3.)

Finding and developing new materials, whether they are *organic*—complex biological molecules with carbon backbones—or inorganic, requires a series of chemical reactions. The experimentation necessary to find the right reactions is time consuming. As with cooks who operate without a recipe, the trial-and-error process with a chemist usually leaves time for the production and testing of only a few hundred compounds. Although this number may seem like a lot, most of the effort is wasted—the product is too sticky, not strong enough, too brittle, too easily damaged by other chemicals, not flexible enough, or has some other fault for which it fails to suit its intended purpose.

But a new technique, beginning in the 1990s, has accelerated the process of generating compounds to be tested. The idea is to start with a chemical or small group of chemicals that the experimenter believes is important—perhaps these chemicals have formed some of the most promising compounds in previous efforts, or they are components of related materials. In the past, chemists needed to mix the chemicals along with other reactants under the proper conditions of temperature, pressure, and so forth, and then use various methods such as *distillation* or *chromatography* to extract the reaction product. To get another product, the experiment must plod through the same steps. But in a technique known as combinatorial chemistry, the chemicals are mixed in a parallel fashion, so that many reactions proceed at the same time.

As described in the following sidebar, the result is a family of related chemicals, ready to be tested.

Combinatorial chemistry produces hundreds or possibly thousands of candidate compounds to screen for the desired properties. One may fit the bill, or come close, in which case it may be further modified and retested. Drug manufacturers often use combinatorial chemistry in conjunction with other methods to create testable candidates for medicinal activity. The section titled "Hitting the Target—Discovering and Designing New Medications" on page 25 describes drug development and design in more detail.

Other scientists besides biomedical researchers have employed combinatorial chemistry. In 1995, X.-D. Xiang and Peter G. Schultz, then at the University of California, Berkeley, and their colleagues published a paper in *Science* describing a pioneering application of this technique in materials research. As described in "A Combinatorial Approach to Materials Discovery," Schultz and his colleagues performed parallel synthesis—a lot of reactions at the same time—to make an ar-

Combinatorial Chemistry

The advantage of combinatorial chemistry is in the large number of combinations that even a small number of chemicals can form. Chemists often start by attaching molecules to tiny plastic beads, or placing them in a solution contained within small wells on a plate, and then begin to add other chemicals. The chemicals may be amines (derived from ammonia, NH_3), hydrocarbons (molecules of carbon and hydrogen), carboxylic acids (containing a COOH group), or many others.

Consider, for instance, chemicals X, Y, and Z. Attach X to the beads, then expose the beads to solutions containing the three chemicals, and allow a reaction to occur. The products will be X-X (chemical X reacted with itself), X-Y, and X-Z. Repeat the process with these products, which is simple and efficient because it is the same process as before, so

ray of different combinations of elements such as bismuth, strontium, barium, yttrium, oxygen, and others. These compounds could then be tested for special electronic properties.

Despite the contribution of methods such as combinatorial chemistry, developing new materials has remained a difficult endeavor. Combinatorial chemistry provides ample candidates to test but offers no clues as to which candidate may be the best.

DESIGNING THE RIGHT STUFF

At the present stage of materials science, researchers either find a good material by chance, in which case they hunt around for legitimate uses for it, or they search for materials with specific properties and use techniques such as combinatorial chemistry to generate a lot of testable compounds. Most of the important materials used today have been found in this way. The process certainly works, for modern technology enjoys an impressive array of materials, including various metals, plastics, ceramics, and

no elaborate preparation is necessary. After the next step, the products will be one of nine types: X-X-X, X-X-Y, X-X-Z, X-Y-X, X-Y-Y, X-Y-Z, X-Z-X, X-Z-Y, and X-Z-Z. If the process is repeated once more, 27 different compounds will appear. Combinatorial chemistry generates a lot of compounds rapidly and efficiently.

The same technique can be used to modify a compound that particularly interests researchers. Suppose this compound almost but not quite fulfills an engineer's requirements. Perhaps after a slight modification, such as the addition of a hydroxyl group (OH) at a certain point or the deletion of an acidic group elsewhere, the compound will work. Researchers may set up conditions in which the compound reacts with a number of chemicals, producing a vast array of compounds that have the same basic structure but differ by a chemical group or two. Somewhere in this collection may be a winner.

composites. But progress has been slow and time consuming. The next stage in materials science, at the frontier of this branch of research, is to investigate the possibility of designing the optimal material in advance.

Designing materials rather than relying on trial and error would be much faster and more efficient. Imagine a team of construction workers trying to raise a skyscraper without benefit of a blueprint or architect—one mistake and the whole thing tips over. Developing new materials without a plan can be just as frustrating.

There are two main design methodologies that researchers are currently pursuing. Both have advantages and disadvantages. One method,

Quantum Mechanics

The British physicist and mathematician Sir Isaac Newton formulated his laws of motion in the 17th century. These laws predict the course of an object when subjected to various forces, such as a push, pull, or a collision with another object. In Newtonian physics, physicists can predict the motion of an object with any desired degree of accuracy if all the forces acting on it are precisely known.

In the 20th century, physicists discovered to their surprise that small particles such as atoms and the components of atoms do not obey Newton's law of motion. Instead of being deterministic—following trajectories determined by the laws of physics—tiny bits of matter behave probabilistically, meaning that their state or trajectory is not precisely determined but can follow one of a number of different options. The German physicist Werner Heisenberg proposed his uncertainty principle in 1927, which states that there is generally some amount of uncertainty in measurements of a particle's state.

Also in the 1920s, the Austrian physicist Erwin Schrödinger developed an equation that describes the motion of small particles. This equation, which continues to be used today, discussed in the present section, involves designing new materials from scratch by applying principles of chemistry and physics. If scientists or engineers gain an understanding of how atoms and molecules impart specific properties to a given material, they could design a new material with a specific property by assembling the essential atoms or molecules; atoms and molecules would be like building blocks out of which the desired material would be made. The other method will be described in the following section.

Quantum mechanics governs the behavior of tiny particles such as atoms. This theory has been confirmed in many different experiments

does not yield the exact values or trajectories but rather probabilities for these quantities. A solution to Schrödinger's equation does not predict the course of a single particle; it assigns probabilities to different courses. For example, 60 percent of the particles in the same situation will follow path A, 30 percent will follow path B, and 10 percent will follow path C.

Some physicists have had trouble accepting the probabilistic nature of quantum mechanics, including some great physicists such as Albert Einstein, who died in 1955 and never embraced this aspect of the theory. Einstein wrote about quantum mechanics in a 1926 letter: "The theory says a lot, but it does not really bring us any closer to the secrets of the Old One [Einstein's name for God or the creator]. I, at any rate, am convinced He does not play dice." Yet as Einstein acknowledged, the formulations of quantum mechanics are accurate. And even though Schrödinger's equation cannot determine exactly which particle does what, the behavior of an aggregate, such as a large piece of matter, is predictable. If, for example, 60 percent of particles behave in a certain way, a researcher can predict with excellent accuracy the behavior of a piece of matter containing thousands or millions of these particles. Since everyday objects consist of billions of atoms, the behavior of these objects appears deterministic and follows Newtonian physics.

since it was formulated in the early decades of the 20th century by physicists such as Max Planck (1858–1947), Niels Bohr (1885–1962), Werner Heisenberg (1901–76), Albert Einstein (1879–1955), and Erwin Schrödinger (1887–1961). But quantum mechanics differs from the much earlier formulations of Sir Isaac Newton (1642–1727), which dominated physics for hundreds of years and continues to be used in many applications. As noted in the sidebar on pages 20–21, large objects such as footballs and planets seem to behave in precisely predictable ways, described by Newtonian physics, but tiny particles such as atoms do not.

Quantum mechanics allows researchers to understand particle behavior on an atomic and molecular level. The equations are accurate but only provide probabilities for the state or path of any given particle, rather than a rigidly determined quantity, as in Newton's laws. Yet if researchers formulate the equations for a collection of interacting particles, such as chunks of matter, the probabilities as given by quantum mechanics describe the overall behavior to a high degree of precision. The problem is that the equations of quantum mechanics are difficult to solve, and when there are many equations, as there must be when formulating the behavior of a group of interacting particles, researchers must use computers to provide solutions. When the equations are coded in programs that run on fast machines, the resulting computer simulation—a computer program that mimics an event or process—can give a detailed picture of the behavior of a piece of matter.

G. Malcolm Stocks, at the Oak Ridge National Laboratory in Tennessee, and Yang Wang at the Pittsburgh Supercomputer Center in Pennsylvania, have used quantum mechanics and computer simulations to analyze the magnetic behavior of small bits of materials such as iron-platinum alloys. The materials and properties these researchers and their colleagues are investigating are related to the storage of information. Computers, certain music players, and other devices store data in binary digital bits—a 1 or 0—the value of which is often governed by the direction of the magnetic field of a small region of magnetized material on a plate or disk. The quantity of information that can be stored in a given size of disk depends on how small can be each 1 or 0 region. Even a modest computer today can store gigabytes (billions of bytes, which are eight-bit chunks of data); yet as video and other data-rich applications have increased—and the physical size of computers has

This computer, Blue Gene/L at the Lawrence Livermore National Laboratory in California, is one of the world's fastest computers and can perform simulations of many physical processes. *(Lawrence Livermore National Laboratory)*

diminished—more data needs to be stored in smaller areas. The problem is that the bits are becoming so small that they contain only a few particles. As described in the sidebar on pages 20–21, while the behavior of large numbers of particles is predictable, individual particles are not. Fluctuations due to temperature could also cause magnetic fields to change and flip between configurations. As a result, computer storage may become unreliable if it is confined in too small a space.

Stocks, Wang, and their colleagues used an extremely powerful computer made by Cray, Inc., called an XT3, to simulate the behavior of a few thousand atoms in the material. This computer, located at the Pittsburgh Supercomputer Center, has 2,048 computer processors to speed up operations; it so fast it is called a supercomputer. Even so, a simulation of 14,400 atoms, including all their important interactions, required 50 hours of computer time. (Since the computer was shared with other users, not all the processors were devoted to this simulation.)

Iron-platinum and similar materials require a lot of energy to alter their magnetic properties, making these substances stable and a good choice for computer storage applications. The researchers discovered that a lot of the fluctuations occur on the boundary of the 14,400-atom block, leaving the interior unaffected. This result suggests than even this small piece of matter, which is on the scale of nanometers, can reliably store information.

PREDICTING THE PROPERTIES

Another design technique being actively explored involves searching for patterns. Whereas the method described in the previous section attempts to understand the behavior of atoms and molecules, this method takes its cue from combinations of these particles. By means of X-ray crystallography and other methods, scientists have already determined the structures of thousands of crystals, composed from a wide variety of elements and compounds. The set of these structures represents an enormous amount of data. Some researchers have begun to use computers to sift through this data, looking for clues as to what elements and compounds produce which structures. With these clues, prediction of the properties of new materials may be possible.

Considering the huge number of compounds and the complexity of materials, this procedure is unlikely to be able to specify the exact structure of a new material in advance. But the clues gained from previously acquired data may narrow the possibilities to a much smaller number, giving material designers a much better handle on what sort of material they require. Since the amount of data is so large, fast computers must be used, and algorithms—computer procedures—must be written to make an appropriate search. Sifting through large data sets for information that is not readily apparent is a process known as data mining.

The Massachusetts Institute of Technology scientists Christopher C. Fischer, Gerbrand Ceder, Kevin J. Tibbetts, and their University of Wisconsin colleague Dane Morgan developed a data-mining algorithm in 2006 to look at crystal structures. The researchers examined a database containing 28,457 structures of binary alloys (which are composed of two metals). With their algorithm, the researchers gathered clues about the possible structures of a given new material. To test the algorithm's accuracy, its predictions could be evaluated in the case of materials whose structure is known but was not in the database.

The initial algorithm suggests possibilities, but Fischer and his colleagues went a little further. They wrote a second algorithm using the principles of quantum mechanics to probe the list of possibilities. By calculating how stable each of the structures might be, this second step identified the most stable structures in terms of energy—these are the ones most likely to occur, for unstable structures tend to evolve into more stable structures. (This is similar to the process of a ball rolling downhill, coming to rest at the lowest state.) As described in the 2006 paper "Predicting Crystal Structure by Merging Data Mining with Quantum Mechanics," published in *Nature Materials,* the researchers succeeded in 90 percent of tests in narrowing down the correct structure to a short list of five possibilities. This five was out of thousands of candidates.

Having algorithms that can predict structures of new or unknown materials gives engineers a better chance of selecting the best material for the job. At this early stage of research, it is not clear how successful materials design can be, or which methodology is better. Perhaps the best technique will employ a range of strategies, as Fischer and his colleagues did by combining data mining with quantum mechanical computations.

New materials for superior computer data storage and other electromagnetic applications often involve alloys, as in the example above. But design strategies also apply to other difficult problems, including the development of new and improved drugs to treat patients.

HITTING THE TARGET—DISCOVERING AND DESIGNING NEW MEDICATIONS

Average life expectancy for Americans jumped from 45 to 50 years in 1900 to almost 80 in the early part of this century. One of the major reasons for the increased longevity was the development of effective antibiotics. Antibiotics are drugs that kill microorganisms but do not affect human cells, so they help the body fend off bacterial and fungal infections. The early part of the 20th century saw the rise of antibiotics such as penicillin and sulfonamide compounds, decreasing fatalities from wound infections and other diseases caused by bacteria or fungi. Many kinds of antibiotic can be found in the modern physician's arsenal.

But new drugs of any type are extremely expensive to develop. The high cost comes from the need to prove their effectiveness—that the drug actually works—and is safe. To avoid injuring people, initial safety tests are done with laboratory animals. These tests cost a lot of money and, because they involve animal experimentation, are objectionable to some people. If the new drug passes these tests, even more money is needed for clinical trials, which establishes the drug's safety in humans and measures its effectiveness in treating the disease. At any stage in the process, a drug can fail, and if so, the money invested in it is wasted. The final cost is a drug manufacturer's version of "sticker shock." In 2003, Joseph A. DiMasi of Tufts University in Massachusetts, Ronald W. Hansen at the University of Rochester in New York, and Henry G. Grabowski at Duke University in North Carolina estimated that a new drug costs about $800 million to develop. Their paper, "The Price of Innovation: New Estimates of Drug Development Costs," was published in the *Journal of Health Economics.*

This expense is a strong motivation for medical researchers to develop a more efficient means of drug discovery than the old trial-and-error method. Rational drug discovery is the name given to techniques that employ the principles of chemistry and physics, or are guided by experimental data, to aid in the search for new drugs.

Drugs commonly work by interacting with certain molecules, which are often proteins. For example, the chemical fluoxetine (which is sold under the name Prozac) inhibits the activity of a protein that pumps serotonin molecules into the cell and is a member of the SSRI class of antidepressants. One of the roles that serotonin plays is to transmit messages between brain cells called neurons, a process that chemists and biologists have been studying since the early 20th century. As researchers discovered the importance of serotonin and similar molecules in brain mechanisms underlying mood and emotion, the possibility of developing drugs that affect this process became clear. This research led to the development of SSRIs in the 1980s, one of the earliest successes in "rational" drug design.

A molecule's structure plays a critical role in its biological activity. Part of the job of many biological molecules is to bind to another molecule—the "target"—and affect its function in various ways. Binding usually requires structural compatibility. This compatibility ensures specificity—the molecule binds only to its target. For instance, enzymes are proteins that speed up specific chemical reactions in the body by binding and positioning the reactants. The shape of the reactants fit with the shape and structure of the enzyme, giving the enzyme its specificity.

Structure-based drug design makes use of the data from X-ray crystallography and NMR techniques. If the biology of the disease process is understood well enough to identify an important component, then researchers can search for a molecule having a structure that might be able to affect this component. For example, protein kinases are a family of enzymes that are involved in many of the functions of a cell. Disruptions of these enzymes have been associated with a significant number of diseases, such as breast *cancer,* and this makes protein kinases an important target for drug development. For this development to succeed, the structure of protein kinases must be studied.

Proteins are composed of a string of amino acids, the sequence of which governs the proteins' properties. In the family of protein kinases, the members share certain sequences—these sequences are "conserved," presumably because they are critical in the function of all protein kinases—and other sequences vary, which provides the specificity of the individual enzymes. Drug designers must identify these sequences and understand their role. In 2007, James D. R. Knight and Rashmi Kothary of the University of Ottawa in Canada, along with Bin Qian and David Baker of the University of Washington in Seattle, accomplished one step in this process by publishing an analysis of protein kinase structures. The paper, "Conservation, Variability and the Modeling of Active Protein Kinases," published in *Public Library of Science (PLoS) ONE,* compared some of the 518 known human protein kinases and introduced an algorithm that may become useful in predicting their structures.

Antibiotics is another area in need of drug development. Although many antibiotics have been developed, microorganisms evolve rapidly. Some of the mutations give microorganisms protection against certain antibiotics, which has caused dangerous and sometimes fatal antibiotic-resistant infections. Ray Zarivach and Natalie Strynadka, at the University of British Columbia in Canada, and their colleagues are studying a number of proteins used by certain bacteria to invade and attack their host. For example, a group of interacting proteins known as the type III secretion system is involved in the injection of bacterial molecules into human cells, which often leads to disease. By studying the structure of these proteins, Zarivach, Strynadka, and their coworkers hope to spark the development of drugs that interfere with this process. In 2006, the researchers published "Structural Analysis of a Prototypical ATPase from the Type III Secretion System" in *Nature Structural & Molecular Biology,* which used X-ray crystallography and other techniques to describe the structure of one of the critical enzymes in this system.

Understanding the structures of both small and large molecules is critical in developing and designing new materials. Speeding up a process that normally involves years or even decades of work—and a lot of frustrating dead ends—will not be easy. The study of protein kinases, bacterial proteins, and many other molecules lays down the foundation on which future drug design methodologies will rest. This work is just beginning. Most medications, as well as materials used in engineering applications, continue to be discovered the old-fashioned way—by relying on good luck or, lacking this, doggedly searching through a horde of candidates until the right one is found. Yet progress is being made in designing materials for a diverse group of applications, from computer components to antibiotics.

The conventional means of reaching space is by rocket. *(NASA)*

CONCLUSION

Designing new chemicals and substances is in its infancy, but its maturation, should this occur, would affect almost every aspect of technology and society. Everything from medicine to construction requires substances having the properties necessary to perform their function. New materials are the crucible from which new technologies emerge; an engineer cannot build a novel device without the appropriate materials.

One of the most ambitious engineering ideas to come along in a while is a project known as the space elevator—a tall structure from which satellites and other objects could be launched. The conventional means of launching payloads is with a rocket propelled by chemical fuel. Rocket launches are risky, and malfunctions are common. They are also tremendously expensive—the cost per pound is about $10,000 ($22,000/kg)! Instead of costly chemical rockets, a space elevator would simply haul the object up to the needed altitude. Having a functional space elevator could reduce this cost by a factor of about 50 or even 100 times.

The difficulty with space elevators is that they would need to be thousands of miles high. A common orbit for satellites is geosynchronous orbit, where the satellite moves at a speed that matches the Earth's rotation. In this situation, the satellite stays in one point of the sky, since its revolution matches that of Earth. The speed of an orbiting satellite depends on its altitude, and the altitude at which the speed of a satellite matches the Earth's rotation—in other words, a geosynchronous orbit—is 22,240 miles (35,784 km). Several designs have been considered for space elevators, with a height reaching 62,000 miles (100,000 km). Compare that with the tallest freestanding (self-supported) structure, Burj Dubai in Dubai, United Arab Emirates, which is still under construction as of May 2009, and is expected to reach a final height of about 2,684 feet (818 m)—slightly more than 0.5 miles (0.8 km).

To reach the fantastic heights required by a space elevator, new materials are essential. Skyscrapers of today are usually composed of a steel framework, but steel is too heavy to use for the space elevator—the tower must be so high and needs so much material that it could not possibly support its own weight if it was made of steel. One possibility for a new material is related to the fullerenes discussed in the text. Carbon nanotubes are cylindrical fullerenes composed of sheets of graphites rolled into tiny cylinders with a diameter of roughly 0.00000004 inches (0.0000001 cm)—a nanometer. The structure and bonds of carbon nanotubes could potentially be used to create a material with about 50–100 times the strength of steel. A slender ribbon of carbon nanotube could be the key to bringing the dream of a space elevator into reality.

A functioning space elevator would drastically change the world, allowing the placement of factories, research laboratories, and even hotels into orbit. Other new materials and medicine could have similar affects, altering the way people live and work in a multitude of interesting ways.

Chemicals and materials are vital components of today's technological world, which has slowly evolved from the use of stone axes and bearskins to the development of satellites, antibiotics, alloys, plastics, and composites. This development is impressive but required thousands of years. If the frontiers of chemistry can expand far enough to include material design, the coming progression of technology will be much more rapid.

CHRONOLOGY

2,500,000 B.C.E.	Human ancestors begin using stone tools by this time.
25,000 B.C.E.	Ceramics are in use by this time.
4,000 B.C.E.	Smelting of copper begins in Egypt and parts of Europe.
3,000 B.C.E.	People in Egypt and Mesopotamia make glass. Mesopotamians discover that a combination—an alloy—of copper and tin produces a strong material, known as bronze.
1,500 B.C.E.	Metalworkers in Greece and parts of Africa and the Near East invent efficient methods to smelt iron.
300 B.C.E.	Chinese metalworkers develop a method of casting iron, producing a high-strength alloy of iron and carbon.
1839 C.E.	American inventor Charles Goodyear (1800–60) develops a process known as vulcanization that makes rubber more durable and elastic, which encourages the widespread usage of this material.
1856	British engineer Sir Henry Bessemer (1813–98) and others perfect an efficient technique to produce steel. This strong combination of iron and carbon

will find enormous use in railroads, shipbuilding, construction, and other vital components of the Industrial Revolution.

1869   Russian chemist Dmitry Mendeleyev (1834–1907) organizes the chemical elements into a periodic table.

1908   Belgian-American chemist Leo Baekeland (1863–1944) makes Bakelite, the first synthetic (artificial) plastic.

1912–14   German physicist Max von Laue (1879–1960) and British scientists Sir William Henry Bragg (1862–1942) and his son, Sir William Lawrence Bragg (1890–1971), discover how to use X-ray diffraction to analyze the structure of molecules.

1938   Russell Games Slayter, a researcher at the American company Owens Corning, perfects a technique to produce fiberglass.

1953   Scientists at ASEA, a Swedish company, create diamonds by using extremely high pressure. A year later, Tracy Hall, a scientist at the American company General Electric, develops a reliable technique to make diamonds (and is often given the credit for generating the first artificial diamond, since the work of the Swedish researchers was not reported until much later).

1985   Harry Kroto, Richard Smalley, Robert Curl, and their colleagues discover a different form of carbon, C_{60}, also known as buckminsterfullerene or "buckyball."

1990s   Combinatorial chemistry comes into use as a technique to produce a large number of compounds quickly and efficiently.

1991	Japanese physicist Sumio Iijima discovers an effective method of producing carbon nanotubes.
2000s	New medications cost hundreds of millions of dollars to develop.
2006	Christopher C. Fischer, Gerbrand Ceder, Kevin J. Tibbetts, and Dane Morgan develop a data-mining algorithm to look at crystal structures.
2007	James D. R. Knight, Rashmi Kothary, Bin Qian, and David Baker describe the structure of many protein kinases, which play an important role in a patient's drug and medication responses.
2009	About half of Boeing's new 787 airliner is made of composite materials.

FURTHER RESOURCES
Print and Internet

Aldersey-Williams, Hugh. *The Most Beautiful Molecule: The Discovery of the Buckyball.* New York: Wiley, 1995. Scientific discoveries are often based on a complicated series of events. This book tells the lively tale of the discovery of buckminsterfullerenes (also known as buckyballs), and discusses their potential applications.

Atkins, Peter. *Atkins' Molecules,* 2nd ed. Cambridge: Cambridge University Press, 2003. Atkins is a chemist and the author of a number of books on science and chemistry. This book describes a variety of interesting substances, including a lot of important biological molecules.

Bessemer, Sir Henry. *Sir Henry Bessemer: An Autobiography.* Available online. URL: http://www.history.rochester.edu/ehp-book/shb/start.htm. Accessed May 28, 2009. Published in 1905, this book offers a look at the prolific inventor and engineer in his own words.

Bleeke, John R., and Regina F. Frey. "Fullerene Science Module." Available online. URL: http://www.chemistry.wustl.edu/~edudev/Fullerene/

fullerene.html. Accessed May 28, 2009. The authors are researchers in the Chemistry Department at Washington University, in St. Louis, Missouri. Topics include the discovery of fullerenes, structure and bonding, synthesis, applications, and more.

Bortz, Fred. *Techno-Matter: The Materials behind the Marvels.* Fairfield, Iowa: 21st Century Books, 2001. Written for young adults, this entertaining book describes many of the advanced materials that form the foundation of much of today's technology.

Chung, Hsiu-Ying, Michelle B. Weinberger, Jonathan B. Levine, Robert W. Cumberland, Abby Kavner, Jenn-Ming Yang, Sarah H. Tolbert, and Richard B. Kaner. "Synthesis of Ultra-Incompressible Superhard Rhenium Diboride at Ambient Pressure." *Science* 316 (April 20, 2007): 436–439. The researchers find a compound that forms short bonds and tight packing, making a hard solid material.

DiMasi, Joseph A., Ronald W. Hansen, and Henry G. Grabowski. "The Price of Innovation: New Estimates of Drug Development Costs." *Journal of Health Economics* 22 (2003): 151–185. The researchers estimate that a new drug cost about $800 million to develop.

Fischer, Christopher C., Kevin J. Tibbetts, Dane Morgan, and Gerbrand Ceder. "Predicting Crystal Structure by Merging Data Mining with Quantum Mechanics." *Nature Materials* 5 (2006): 641–646. The researchers developed a data-mining algorithm in 2006 to look at crystal structures.

Gordon, J. E. *The New Science of Strong Materials, or Why You Don't Fall through the Floor.* Princeton, N.J.: Princeton University Press, 2006. This revised edition of a science classic, written by an expert in the field, explains the physics and chemistry of materials.

Knight, James D. R., Bin Qian, David Baker, and Rashmi Kothary. "Conservation, Variability and the Modeling of Active Protein Kinases." *Public Library of Science (PLoS) ONE,* October 3, 2007. Available online. URL: http://www.plosone.org/article/info%3Adoi%2F10.1371%2Fjournal.pone.0000982. Accessed May 28, 2009. The researchers compared some of the 518 known human protein kinases, and introduced an algorithm that may become useful in predicting their structures.

Minerals, Metals, & Materials Society. "Greatest Moments in Materials Science and Engineering." Available online. URL: http://

www.materialmoments.org/top100.html. Accessed May 28, 2009. Presentation of the top 100 developments in the long history of materials science, as compiled in a survey conducted by *JOM,* the journal of the Minerals, Metals, & Materials Society.

Read, Randy J. "Overview of Macromolecular X-ray Crystallography." Available online. URL: http://www-structmed.cimr.cam.ac.uk/Course/Overview/Overview.html. Accessed May 28, 2009. This Web page offers a well-illustrated description of the principles and practice of X-ray crystallography, an extremely important technique in determining the structure of materials.

University of Illinois Urbana-Champaign. "Materials Science and Technology." Available online. URL: http://matse1.mse.uiuc.edu/~tw/. Accessed May 28, 2009. Thanks to the hard work of dozens of high school teachers, as well as a number of college professors and students, this Web site presents a highly informative set of pages on the science of materials. There are modules on ceramics, metals, polymers (plastics), composites, concrete, and more.

Xiang, X.-D., Xiaodong Sun, Gabriel Briceño. Yulin Lou, Kai-An Wang, Hauyee Chang, William G. Wallace-Freedman, et al. "A Combinatorial Approach to Materials Discovery." *Science 268* (June 23, 1995): 1,738–1,740. The researchers performed parallel synthesis to make an array of different combinations of elements such as bismuth, strontium, barium, yttrium, oxygen, and others.

Zarivach, Raz, Marija Vuckovic, Wanyin Deng, B. Brett Finlay, and Natalie C. Strynadka. "Structural Analysis of a Prototypical ATPase from the Type III Secretion System." *Nature Structural and Molecular Biology* 14 (2007): 131–137. The researchers used X-ray crystallography and other techniques to describe the structure of a critical enzyme.

Web Sites

Chemical Abstracts Service. Available online. URL: http://www.cas.org/. Accessed May 28, 2009. A division of the American Chemical Society, the Chemical Abstracts Service has one of the most extensive collections of chemical information in the world.

Lawrence Livermore National Laboratory. Available online. URL: http://www.llnl.gov/. Accessed May 28, 2009. This Web site contains news and information on the research conducted at the laboratory. Especially useful is the online version of *Science & Technology Review,* a magazine published 10 times a year by the laboratory. Each issue contains interesting and accessible articles about some of the many ongoing research projects at Lawrence Livermore.

Los Alamos National Laboratory: Periodic Table of Elements. Available online. URL: http://periodic.lanl.gov/. Accessed May 28, 2009. This table contains clickable elements that lead to a Web page describing the element's history, properties, sources, uses, and other information.

Pittsburgh Supercomputer Center. Available online. URL: http://www.psc.edu/. Accessed May 28, 2009. This Web site provides news and information about the computers and projects of the Pittsburgh Supercomputer Center. These computers are used by researchers at the center as well as by scientists at universities and other research organizations for analysis, simulation, and prediction in a broad variety of scientific disciplines, including materials science. The Web site contains an online version of the annual magazine *Projects in Scientific Computing,* published by the Pittsburgh Supercomputer Center, which details some of the research highlights each year.

NANOTECHNOLOGY— TECHNOLOGY ON A MOLECULAR SCALE

The Dutch craftsman Antoni van Leeuwenhoek (1632–1723) built some of the finest microscopes available in the 17th century. The quality of a microscope depends on the quality of the lens or lenses it uses to bend light and form an image. Leeuwenhoek's microscopes contained a single lens, and this lens was so carefully made and ground into the correct shape that his instruments were superior to most others. When he used the microscope to peer into drops of water, Leeuwenhoek was amazed to see tiny swimming creatures.

Leeuwenhoek and other early microscopists unveiled a new branch of science with the discovery of microorganisms and other tiny objects, invisible to the unaided eye. Yet scientists of the 17th and 18th centuries did not understand this new, miniaturized world. They had a tendency to describe and think about these objects as if they were tiny replicas of larger, more familiar objects. For instance, Leeuwenhoek and fellow Dutchman Nicolas Hartsoeker (1656–1725) took this attitude when they studied male reproductive cells called semen. These little cells, when united with female reproductive cells called ova, initiate an amazing and complex developmental process by which the cells grow, divide, and become a human being. But this process was unknown in the 17th century. When Leeuwenhoek and Hartsoeker examined semen, they thought they saw a tiny version of a human being, waiting for the proper moment to pop out and start growing. Leeuwenhoek referred to it as a little man.

The little man, or homunculus, as it was sometimes called, was imaginative, of course—the mind sometimes creates such illusions when examining an object too small or fuzzy to be clearly seen. The notion that the microscopic world consisted of familiar objects reduced to a small scale was also incorrect. As scientists probed the nature of particles, they realized that the behavior of small objects does not necessarily mimic larger ones. This discovery opened up new vistas in science as well as technology, including the subject of this chapter—technology on the scale of atoms and molecules.

INTRODUCTION

People have been thinking about tiny objects for a long time. The ancient Greek philosopher Democritus (ca. 460–370 B.C.E.) believed that properties of matter depended on the shapes of small, indivisible bits of matter called atoms. Although this idea failed to catch on at the time— no one could see these atoms because they were so small—in 1803, the British chemist John Dalton (1766–1844) proposed a similar theory. Dalton's theory was an important advance and helped scientists understand chemical reactions—for example, the reaction of two atoms of hydrogen (H) and one atom of oxygen (O) to form H_2O—but atoms themselves remained cloaked in mystery.

The Scottish physicist James Clerk Maxwell (1831–79) imagined in 1867 a little being that was small enough to see and manipulate molecules. Maxwell used his imaginative creation to think and hypothesize about the laws of *thermodynamics*—a branch of physics that deals with heat and the motion of molecules—but the fictional creature, later known as Maxwell's demon, also encouraged people to think about how to control or manipulate atoms, which was an exciting prospect. People began developing processes that involved small particles, although in an indirect manner. Chemists set up reactions with specific atoms and molecules, but of course these reactions involved matter in bulk quantities.

An early application of small quantities of matter occurred in photography. A thin film or glass plate, when coated with tiny particles of

Circuit board containing integrated circuits *(Stefan Witas/ iStockphoto)*

silver compounds such as silver bromide, formed an image in the presence of light. Another application involved charged particles known as electrons, which formed beams that could be used to strike a phosphorus screen and produce light. This process became the basis for the cathode ray tube (CRT) and early television sets, and is sometimes still used for this purpose. The electron, along with protons and neutrons, are components of atoms, and their isolation proved that atoms are not actually indivisible, as had been earlier believed.

In 1958, Jack Kilby (1923–2005), a researcher at Texas Instruments, invented a miniature electronic circuit known as an integrated circuit. Modern integrated circuits, which contain many electronic components etched on a thin wafer of silicon, have become a vital part of many devices, especially computers. Placing a large number of electronic components in such a small area has greatly reduced the size of electronic equipment.

But even more ambitious goals emerged. The American physicist Richard Feynman (1918–88) gave a lecture in 1959, "There Is Plenty of Room at the Bottom," at a meeting of the American Physical Society. Feynman's lecture discussed controlling and manipulating objects on a small scale, and he began by wondering what kind of technology could write all 24 volumes of the *Encyclopaedia Britannica* on the head of a pin.

Believing that no law of physics prohibits small-scale technologies, Feynman argued, "I am not inventing anti-gravity, which is possible someday only if the laws are not what we think." In contrast to something that violates the laws of physics as they are currently understood, decreasing the size of devices was feasible. In other words, in Feynman's creative language, people had room with which to work at the low end (bottom) of the size scale, and the reason nobody had done it was "simply because we haven't yet gotten around to it." A highly respected physicist, Feynman won a share in the 1965 Nobel Prize in physics for his work on applying quantum mechanics to electromagnetic phenomena.

Feynman's idea was a little ahead of its time, but in 1986 the American engineer Eric Drexler published a book, *Engines of Creation,* in which he popularized the development of small-scale technologies. Drexler had been influenced by Feynman's lecture but took the idea further, sparking the imagination and curiosity of other engineers and scientists. People had been thinking about this subject earlier—the Japanese physicist Norio Taniguchi coined the term *nanotechnology* in 1974—but Drexler's enthusiasm was important in calling attention to the field.

Nanotechnology takes its name from the nanometer, which is a billionth of a meter. (*Nanos* is a Greek word meaning dwarf or midget.) A nanometer is about 0.00000004 inches (0.0000001 cm). It is difficult to visualize such a tiny length. Imagine taking a piece of paper one inch (2.5 cm) long and dividing it into 25,000,000 segments, each segment of which will be a nanometer. This length is much smaller than anything people experience in everyday life. A gnat, for instance, is about 0.1 inches (0.3 cm), and the diameter of a human hair is roughly 30 times smaller. Bacteria are 10–50 times smaller than the diameter of a human hair, and a nanometer is 5,000 times smaller than the average size of a bacterium. Roughly 10 atoms would fit along the length of a nanometer.

In the most common definition, nanotechnology refers to any technology or device on the scale of about 1–100 nanometers. Microscopes that use light are not useful in this range because the wavelength of visible light exceeds these magnitudes, and distortion due to diffraction—the scattering of light rays—obscures the image of objects of this size.

Researchers turn instead to electrons. The German engineers Ernst Ruska (1906–88) and Max Knoll (1897–1969) invented the electron microscope in the 1930s; this microscope creates images by passing a beam of electrons through a sample, or sometimes reflecting electrons from the surface. Although it seems odd, electrons have a wavelength—in the strange world of quantum mechanics, all objects have both particle and wave properties. Electrons have tiny wavelengths, giving electron microscopes the power to produce images on the scale of nanotechnology. Disadvantages of electron microscopes include high cost and the need for the sample to be placed in a vacuum since air molecules disrupt the electron beam. But the magnification of electron microscopes is unattainable with their light-based counterparts.

With electron microscopes providing the "eyes," researchers explore the "nanoworld." Miniaturization of electronics has reached the point where components of an integrated circuit can be as small as 0.000002 inches (50 nm), well within the scale of nanotechnology. Even more ambitious developments are on the drawing board or in the laboratory. As described below, some researchers are hoping to use atoms as building blocks to construct motors on a vanishingly small scale.

But science and engineering on the scale of nanotechnology is much different from manipulating ordinary, macroscopic projects. It would be relatively simple to take a design for a large object and make it smaller by reducing its dimensions—for instance, instead of making a box having three inches (7.5 cm) on each side, the manufacturer would make the box a million times smaller.

Reducing each component would, in principle, create a miniaturized instrument or device, but in most cases, this simple idea probably would not work. Quantum mechanics becomes important at the scale of nanotechnology, and as described in the sidebar on page 20–21, its laws are different from Newtonian physics.

But there are other problems with the notion that the nanoworld is just a smaller version of the everyday, *macroscopic* world—the whole environment in the nanoworld is different. Consider gravity, for instance. Gravity is enormously influential in the macroscopic world because large pieces of matter exert strong gravitational forces. But on the scale of nanotechnology, everything is so tiny that the force of gravitation is ignorable. Imagine two astronauts, floating in space. Each astronaut has mass and would exert a gravitational attraction on the other, but the astronauts are so small that the force would be negligible. They could easily separate—for example, by pushing against each other, or by using the thrust of an exceedingly small jet pack.

Instead of gravity, electromagnetic forces are the major "players" in the nanoworld. This is a different environment, which along with the effects of quantum mechanics, means that materials can have different properties in nanotechnology than they have in the macroscopic world. As discussed in the remainder of the chapter, this can have advantages as well as disadvantages.

As in all active areas of research, there are several approaches to the development of nanotechnology. One approach involves techniques similar to sculpting, where one starts with a large piece of material and cuts away what is not needed. The problem is that a lot of the material winds up wasted on the cutting room floor. An alternative approach, described in the following section, starts at the bottom or lower end of the scale and builds up from there. An important feature of this approach is the bonding of molecules to make even larger, more complex molecules—supramolecular chemistry.

SUPRAMOLECULAR CHEMISTRY

The previous chapter discussed covalent and ionic bonds. These are strong chemical bonds formed when atoms share electrons, in the case of covalent bonds, or transfer electrons, as in ionic bonds. A large portion of chemistry textbooks are devoted to these chemical bonds, and rightfully so, considering their importance in the formation of many of the compounds and materials that people use. But there are other kinds of bonds, generally referred to as noncovalent bonds, that are weaker and often temporary, breaking and forming repeatedly. These bonds play a vital role in certain materials and are the primary mechanisms involved in supramolecular chemistry.

Many biological molecules, such as proteins, are large and have complex structures. Proteins are composed of a sequence of units, called amino acids, which are joined by a strong covalent bond known as a peptide bond. But proteins also fold up into a certain shape, which is vital to their function—a protein that loses its shape cannot fulfill its function as an enzyme or a transporter. The sequence of amino acids determines the shape; 20 different amino acids are found in proteins, and each one has slightly different chemical and physical properties. For example, some amino acids are hydrophilic, readily interacting

Hydrogen Bonds

Hydrogen is the smallest atom, consisting of a single proton and electron. (Some *isotopes* of hydrogen contain one or two neutrons in the nucleus, but the vast majority of hydrogen atoms have no neutrons.) Covalent bonds form when atoms share electrons, which stabilize their electron configurations. When a hydrogen atom forms a covalent bond with certain atoms such as oxygen, fluorine, and nitrogen, it shares a pair of electrons with the other atom—one from the hydrogen and one from the other atom. But the sharing of the electron pair is not an equal, 50-50 partnership. Oxygen, fluorine, and nitrogen are strongly electronegative, meaning that these atoms tug the "shared" electrons more to their side of the bond. Tiny hydrogen cannot resist this pull. The result is that the negative charge of the electrons tends to be more toward hydrogen's partner, leaving the positively charged proton of hydrogen "uncovered." This process creates electric fields that attract opposite fields and repels like fields. As shown in the figure at right, the hydrogen end of the molecule, with its partial positive charge, sticks to oxygen, fluorine, and nitrogen ends of other molecules.

A hydrogen bond is typically about 20 times weaker than a covalent bond. But hydrogen bonds are strong enough to draw molecules together, forming attachments that can be temporary but have important effects. Water molecules form plenty of hydrogen bonds, for example, and as a result, the molecules stick together more readily than other molecules. This is why water remains a liquid even at high temperatures—heat drives molecules apart, but more heat is needed to break the hydrogen bonds holding water molecules together. A similar compound, hydrogen sulfide (H_2S), does not tend to form hydrogen bonds and has a boiling point of –77°F (–60°C), which is much lower than the boiling point of water, 212°F (100°C). Hydrogen sulfide is a gas well below room temperature.

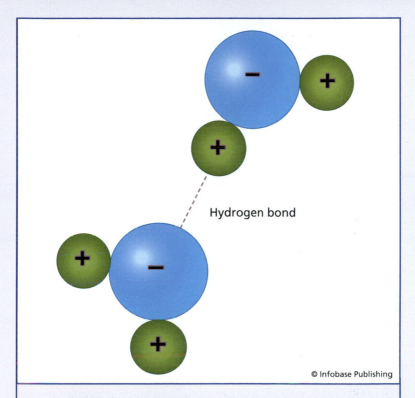

© Infobase Publishing

The unequal sharing of electrons, such as in between oxygen (represented in the figure as a large circle) and hydrogen (small circle), produces electric fields. In this example, the negative field of the oxygen atom of one water molecule attracts the positive field of a hydrogen atom of another water molecule, thereby forming a relatively weak bond known as a hydrogen bond.

DNA is another molecule in which hydrogen bonds have a vital role. This long molecule consists of sequences of units known as bases, which are joined by strong chemical bonds. Two strands wrap around each other in a helix, and hydrogen

(continues)

(continued)

bonds between the bases hold the strands together—these bonds stabilize DNA, maintaining its structure for the life-time of the cell (otherwise, the cell will die). Certain sequences encode genes, but before these sequences can be read, the strands must be pulled apart to expose the bases. If the strands were connected by strong bonds, pulling them apart would be a major operation. With weak hydrogen bonds, the strands can be repeatedly separated and then reconnected, allowing the cell to read out the genes but maintaining the stability and longevity of its DNA.

with water molecules, and some are hydrophobic, strongly avoiding water molecules. One of the major factors governing the shape of most proteins is that they usually function in aqueous solution and fold in such a way that hides hydrophobic amino acids on the inside.

Other molecular forces are electrical. Large molecules often have charged groups attached at certain points, and these charges exert electrical forces on one another as well as on charged molecules in the surroundings. One of the most important of the weaker bonds is electrical in nature and arises because of unequal charge distribution. This bond, usually called a hydrogen bond, is important in a lot of molecular structures, including protein folding and the joining of two strands of a double helix of deoxyribonucleic acid (DNA). As described in the sidebar on pages 42–44, this bond is known as a hydrogen bond because one of the participants is a hydrogen atom.

Creating novel molecules is a process of finding the right building blocks and the conditions under which they form bonds. Chemists often synthesize substances by setting up a chemical reaction or a series of reactions whose final product is the desired substance; supramolecular chemistry is similar, but generally it involves large molecules and non-covalent bonds.

Under ideal circumstances, the component molecules will self-assemble—in which case chemists do not have to perform an elaborate series of steps to encourage the union. Self-assembly occurs when the components are more stable together than they are apart. In other words, the assembly reduces the energy so that it happens spontaneously, similar to a spontaneous chemical reaction such as a pellet of sodium reacting violently when immersed in water. Spontaneous processes result in more disorder—this is a law of thermodynamics—and self-assembly would, at first glance, seem to violate this principle. But the principle is upheld because these processes raise the disorder of the overall environment. This is also what happens in living organisms, which is why life fails to violate the laws of thermodynamics even as it grows, evolves, and sustains itself spontaneously.

The job of the chemist is to set up the necessary conditions. Hydrophobic forces can be involved, and the researcher Eugene Zubarev at Rice University in Houston, Texas, and his colleagues have made *nanoparticles* of gold with the aid of these forces. (The size of a nanoparticle is in the nanotechnology range—0.00000004–0.000004 inches [1–100 nm].) Zubarev attaches a tiny gold particle to an amphiphilic molecule—this molecule has both hydrophobic and hydrophilic ends. When placed in water, the amphiphilic molecules spontaneously form cylinders or spheres as their hydrophobic portions bunch together, avoiding water, while the hydrophilic portions are exposed. The same thing occurs as proteins fold, as well as in cellular membranes—these membranes are composed of amphiphilic molecules that form a two-layered structure that enclose cells.

Zubarev's assembly is so biologically adept that it travels widely in the body, and by attaching other molecules such as medications to the gold particle, the assembly can transport substances to specific points in a patient. Pinpoint delivery of drugs would vastly improve treatment regimens because it would tend to avoid side effects, which commonly occur when drugs act on tissues other than the intended target.

Other researchers are employing DNA to promote nanoparticle assemblies. As mentioned in the sidebar on pages 42–44, hydrogen bonds knit the strands of the DNA's double helix. DNA consists of strings of four bases: adenine (A), thymine (T), cytosine (C), and guanine (G). The structures of the bases and the helix permit only the binding of A with T, and C with G. This is called complementarity, for if the base at one location is an A, the base on the opposite strand will be its complement, T, not a C or a G. If it is a G, then the other base will be a C.

Mathew Maye, Oleg Gang, and their colleagues at Brookhaven National Laboratory in New York use DNA to help assemble nanoparticles.

Strands of DNA self-assemble if they are complementary. For example, a strand with the sequence AAGCT is complementary to TTCGA, and will readily join via hydrogen bonds if mixed together in the same solution. The researchers create nanoparticles by attaching the components to complementary strands of DNA. By customizing the sequences and the positions of the attachments, Maye, Gang, and their coworkers can maneuver the components in a variety of ways to form clusters or aggregates. Some of this research has been published in a 2006 paper, "A Simple Method for Kinetic Control of DNA-Induced Nanoparticle Assembly," in the *Journal of the American Chemical Society*.

Nanoparticles can be employed in transportation, as mentioned for Zubarev's assembly, as well as becoming components for larger assemblies. Some of these molecular assemblies can even function as miniature motors.

MOLECULAR MACHINES

Examples of naturally occurring molecular machines and motors are already known. Proteins are the workhorses of cellular biology, performing functions such as transportation, communication, and structural maintenance. Some of these proteins have moving parts and act exactly like small machines.

Two examples are dynein and kinesin. These huge proteins latch onto vesicles—membrane-bound compartments—filled with certain molecules and transport this cargo to another region of the cell. The proteins "walk" along filaments inside the cell, using the energy-rich compound adenosine triphosphate (ATP) as fuel.

Another example is the enzyme that winds apart the two strands of DNA. The hydrogen bonds hold the two strands of the double helix together. Cellular mechanisms need to break apart those strands on occasion to read a gene or to replicate its DNA, and they use enzymes to perform this operation. The enzyme that speeds up the unwinding of the strands is known as helicase. Helicase binds to DNA, and under certain conditions the binding alone might provide enough leverage to break the bonds. But Daniel S. Johnson, Michelle Wang, and their colleagues at Cornell University in Ithaca, New York, discovered that helicase does a little more—not only does it bind, it also yanks the strands apart by exerting force. Wang and her research team made this finding by attaching a tiny plastic bead to one strand of a double helix and anchoring the other strand to a firm surface. By using a laser beam, the experimenters made precise measurements of the movement of the bead, observing the forces imposed by helicase enzymes. The report of this research, "Single-Molecule Studies Reveal Dynamics of DNA Unwinding by the Ring-Shaped T7 Helicase," was published in 2007 in *Cell*.

Scientists study protein motors because they are biologically interesting but also because they offer insights into mini-motors. Molecular machinery such as nanorobots or nanobots—tiny robots—is a major goal of nanotechnology, and it would have tremendous applications in a lot of fields, especially in medicine. Some researchers are trying to adapt protein motors to perform additional jobs, while other researchers simply use these tiny motors for inspiration. Ever since the 1966 film *Fantastic Voyage*, in which scientists shrank a team of specialists and a submarine and injected them into the body of a patient, people have been fascinated with potential treatments that would be made possible by tiny machines.

Alex Zettl and his colleagues at the University of California, Berkeley, have been active in this field. In 2005, Zettl and his team built an oscillator that is one of the smallest electric motors in the world. The motor works by driving some of the atoms of a tiny drop of liquid metal over to an adjacent and smaller drop. An electric current powers this movement. The most interesting phase of the operation occurs during the rebound, as the atoms return to the original source. This phase happens because of a molecular force known as surface tension.

Forces in the nanoworld can appear strange because they do not exist in the same proportion as forces in the macroscopic world. Gravitational forces are insignificant to molecules because this force is weak unless large masses are involved. Electrical interactions, as in the electric current driving the atomic movement in Zettl's motor, are tremendously more important than gravity on this scale.

Interactions such as surface tension also become more influential at smaller scales. Surface tension is due to attractions between molecules, such as electrical charges and hydrogen bonds, which cause them to bunch together and resist separation. For example, such forces create a "tension" on the surface of water because the molecules stick together, and this force is strong enough to let a bug known as the water strider walk on water. In Zettl's motor, the smaller drop grows until it reaches a radius of about 0.0000012 inches (0.000003 cm or 30 nm), about three times less than the larger drop. The curvature of the smaller drop is greater than the larger drop, which means the smaller drop has greater tension; when the droplet surfaces touch, this force quickly pushes atoms back to the larger drop. At this instant, the oscillator "relaxes" into its initial state. Electricity then begins to drive them back to the smaller drop, starting the oscillation cycle again.

At such small scales, the experimenters cannot see the motor working by any means except an electron microscope. Although the motor is simple conceptually, its precision is incredible—it operates at the atomic level, controlling the motion of atoms as they shuffle back and forth between nanoparticles. B. C. Regan, Zettl, and their colleagues published the report "Surface-Tension-Driven Nanoelectromechanical Relaxation Oscillator" in *Applied Physics Letters* in 2005. As the researchers note in their report, "[S]urface tension can be a dominant force for small systems," as illustrated in their motor. This is a prime example of the different forces and situations that must be taken into account in the nanoworld.

Tiny machines such as Zettl's oscillator may be useful on their own but also in forming the components of more sophisticated instruments such as nanobots. Robotic automation is commonly employed in industrial factories to do jobs that require repetition or extreme precision, such as spot welding. Nanobots would have the added benefit of being able to function in otherwise inaccessibly small locations. Tasks for nanobots include scanning a load-bearing surface and looking for signs of structural failure that would be impossible for a human inspector to see.

A large number of nanobots would be required for most jobs, and that number would probably vary considerably depending on workload. Manufacturing this army of molecular machines might become prohibitively expensive unless the nanobots can self-assemble. An ideal strategy would have self-replicating nanobots that automatically replace damaged or lost machines and ramp up production when an especially demanding task arises.

But the idea of self-replicating nanobots makes some people uneasy. It is not difficult to imagine a scenario in which the machines begin to replicate out of control; similar catastrophes occurring in the human body are called cancers—mutations cause a cell to keep growing and dividing, producing an abnormal growth that can endanger the person's life. In the worst case for nanotechnology, uncontrollable machines might proliferate and consume all resources, making a kind of "gray goo" that covers the world.

The current state of nanotechnology is not nearly advanced enough to pose such extreme threats. Yet it is prudent to plan ahead. The progress of nanotechnology since Richard Feynman delivered his lecture has been rapid and may quickly attain a point at which tremendous benefits—and risks—are realized. Organizations such as the Center for Responsible Nanotechnology, a nonprofit research and advocacy group, are investing time and thought into approaches that are balanced in terms of risk and reward. The "gray-goo" scenario is unlikely because any self-replicating machine would be sensitive to its environment—as are living organisms—which would make it impossible for the machine to expand beyond certain boundaries. But the potential threat of any new technology should be considered along with its benefits.

BUILDING WITH ATOMS

Zettl's oscillator exerts control over the motion of a relatively small number of atoms, but the goal of some nanotechnology researchers is to manipulate atoms themselves. Thanks to the development of the Scanning Tunneling Microscope (STM) in 1981, this once unthinkable feat is not only possible, it has been performed.

As described in the sidebar on pages 50–52, the principles of quantum mechanics are involved in the STM's operation. Considering the scale of operation, this is unavoidable. Although quantum mechanics consists of some unfamiliar concepts, the equations provide an accurate and reliable way of understanding behavior on the atomic level.

Although STMs do a splendid job of providing atomic-scale images, nanotechnology specialists are also interested in the machine's ability to move a single atom, as mentioned in the sidebar on pages 50–52. In 1989, the IBM scientist Don Eigler discovered that he could move around atoms with the STM. He spelled the company letters, IBM, with xenon atoms on a nickel surface. Since then, scientists have yielded to a creative impulse and made a variety of artistic pictures with small numbers of atoms. (Such art not only requires an STM to make, it also requires one in order for people to see and appreciate.)

Besides creating interesting pictures, one of the projects that interests IBM is the use of STM to make even smaller electronic circuits.

Scanning Tunneling Microscope

According to quantum mechanics, atoms do not tend to have an exact position and velocity. Due to variations in motion, atoms can jump around, even to the extent of escaping a container or passing through a barrier. This behavior is not observed in everyday activities—macroscopic objects contain so many atoms that the fluctuations of a few individuals are unimportant, and, as described in the sidebar on pages 20–21, the behavior of the group is predictable. But quantum mechanical effects appear in objects the size of atoms. A tennis ball will always bounce back when it hits the net, but an atom may occasionally encounter a barrier and continue through. Passage through a barrier is called tunneling in quantum mechanics.

Developed in 1981 by Gerd Binnig and Heinrich Rohrer at the IBM Zurich Research Laboratory in Switzerland, the Scanning Tunneling Microscope passes an extremely thin

Researcher using a scanning tunneling microscope *(Volker Steger/ Peter Arnold, Inc.)*

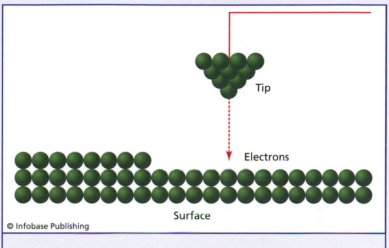

Tip

Electrons

Surface

© Infobase Publishing

Electrons tunnel between the gap formed by the STM probe tip and the surface of the material being scanned. The amount of current formed by the tunneling electrons depends on the size of the gap, and as the probe glides above the surface, it creates an atomic scale map.

probe over the surface to be scanned. As illustrated in the figure above, the tip of the probe tapers to a single atom. (These tips are easily made by etching or tearing a filament of metal.) The machine holds the probe tip above the surface at a height of about one atomic diameter. A small voltage, usually about a volt, is applied to the tip, and a current consisting of electrons begins to flow. A large voltage would produce a current by ripping electrons from the surface, but such a violent event would also ruin the material; a single volt cannot normally pull electrons across the barrier, but it encourages some electrons to "tunnel" to the other side. This current is extremely dependent on distance. The magnitude of current can be used to indicate distance, or the probe tip can be moved up or down to maintain a steady current, in which case the distance to the surface is measured by the probe movement. In either case, the

(continues)

(continued)

machine produces a contour of the surface with the precision of the size of the probe tip—a single atom!

When researchers position the tip even closer to the surface, sometimes an atom will stick to the probe. If this attractive force is strong enough, the atom will break free of the surface and follow the probe. By picking up an atom and then placing it down at another spot, STM allows scientists to move material one atom at a time.

Semiconductors are made by "doping"—adding a small quantity of boron or phosphorus to silicon; controlling this process at the atomic level might allow engineers to construct integrated circuits on a scale unimaginably small prior to the development of STM. The decrease in the size of the components of an integrated circuit means that more of them can be used, resulting in even more powerful computer processors.

NANOMATERIALS

Building materials atom-by-atom is not generally feasible yet, despite the help of machines such as STMs. A square nanometer box—each side having a length of 0.00000004 inches (0.0000001 cm), or, in other words, one nanometer—contains about 500–1,000 atoms, many more particles than researchers can efficiently move at the present stage of science. But nanotechnology is not solely about moving atoms, designing self-assembling molecules, and making tiny motors. Nanotechnology also involves small-scale materials—"nanomaterials."

Assembly is a field of nanotechnology that starts at the bottom, to use Feynman's expression. The molecular construction proceeds by putting together the pieces, either through a series of chemical reactions performed by a chemist or through a process of self-assembly. A different approach to nanotechnology begins with a material of interest. The challenge is to fashion an extremely small yet functional object or particle—a nanoparticle, to use a term common in nanotechnology—out of that material. In practice, this approach may involve whittling, crushing, tearing, or etching a material, though the process often relies on more subtle manipulations.

What are the advantages of smaller sizes? Saving space is not the only advantage, but it is perhaps the easiest one to appreciate. Tiny objects can fit in places that other objects cannot go, and in electronics, packing an increasing number of components on integrated circuits used in computers produces a faster, more powerful machine.

Another advantage is related to physical chemistry. Many important events and chemical reactions occur on the exposed surfaces of a material since the surface is where other objects can easily reach and

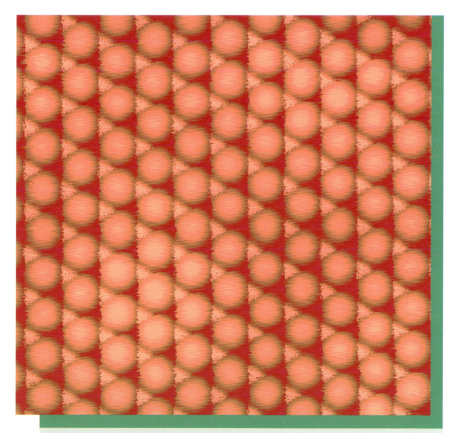

STM image of a tightly packed lattice of copper atoms *(Joseph Stroscio; Robert Celotta/NIST)*

interact with the material. A small, thin object exposes more surface area than the same amount of matter compressed into a thick block, and the greater exposure speeds up interactions with the environment and other objects. For example, crushed ice melts faster than the same mass of ice cubes—the smaller, thinner particles of the crushed ice have more surface area with which to absorb heat, so they require less time to melt.

Nanoparticles have an exceptionally high ratio of surface area to volume—unlike bulky forms of matter, most of the material in nanoparticles is exposed rather than hidden inside the interior. This feature of nanoparticles can drastically alter the material's properties, which is another example of the changes that occur at the small end of the scale. Gold, for instance, is usually not very chemically active in its bulk form, yet gold nanoparticles are quite reactive. An example of this phenomenon will be described in the following section.

Manufacturers have already begun to take advantage of some of these nanoparticle properties. Sunscreens, which protect users from burns by absorbing or deflecting harmful rays, are often made from chemicals such as titanium dioxide or zinc oxide that are particularly effective. These sunscreens often leave a whitish residue—which used to be common on the nose of a pool or beach lifeguard—but when companies embedded nanoparticles of titanium dioxide or zinc oxide instead of bulkier particles, the creams become transparent yet maintained or even increased their effectiveness. With no embarrassing residue, these sunscreens have become popular.

But some health and environmental groups are concerned over such uses of nanoparticles. New technologies may introduce new risks as well as new benefits. The Food and Drug Administration (FDA), the U.S. government agency that oversees drugs and health-care products such as sunscreens, had already approved the use of titanium dioxide and zinc oxide, and the administration decided in 1996 that the use of tiny particles of these chemicals in sunscreens and other health or beauty products did not pose any additional risks. But in 2007, Friends of the Earth, a group based in Washington, D.C., urged a ban on this use of nanoparticles until more research has been done. Proponents of such a ban argue that the novel properties of nanotechnology materials make them behave like new substances and therefore require a new round of tests even though the chemicals themselves have already been approved.

Other nanomaterials, such as those being developed for electronics or industry, have raised fewer health concerns. A prominent field of research involves extremely small crystals that are sometimes called "nanocrystals." Crystals are solids in which the constituent atoms or molecules are packed in repeating geometric patterns—think of a lattice composed of cubes, hexagons, or some other structure. Structure is important, and a different arrangement of components can bestow properties as different as diamond and graphite, which as discussed in the previous chapter are both made of carbon but have different structures. Some crystals are tremendously strong; spider silk is thin and lightweight but strong enough to hold a struggling fly (and seriously annoy a person who blunders into the web), and is reinforced with small-scale crystal material. Spider silk is even stronger than the same weight of steel.

Nanocrystals of silicon are particularly interesting. Silicon is widely used in the electronic industry as a *semiconductor*—it conducts an electrical current under certain conditions. Most metals are excellent conductors because they have mobile electrons to carry the current, but semiconductor materials have few electrons with sufficient energy to do so. Applying a voltage in a certain way to a semiconductor kicks more electrons into action and offers a controllable way to turn currents on or off as needed. Semiconductors make excellent switches, vital in the information-processing circuits of a computer and in many other devices.

When semiconductors such as silicon are sliced into sizes as small as nanoparticles, their electrons adopt unusual properties. Electrons in a large, bulky semiconductor are numerous and have a variety of energy levels that seem to form a continuous range, all bunched together. This is true even though one of the fundamental principles of quantum mechanics is that particles have discrete instead of continuous energy levels—discrete means the energy quantities are distinct and separated, similar to the integers 1, 2, and 3, rather than continuous quantities such as the values of the real number line. The reason electrons in bulky solids seem to have a continuous energy range is that the crowded particles interact, which smears out the discrete energy levels. But when the number of electrons is small, as in a nanoparticle, they have more elbow room, so to speak, and the discrete nature of their energy levels becomes evident. This is a quantum mechanical effect, and the tiny piece of material is accordingly called a *quantum dot*.

Electrons can emit light when they fall from a higher energy level to a lower one. A light-emitting diode (LED) is a semiconductor in which the electrons emit light of a certain wavelength (color). Bulky materials tend to emit light of a fixed wavelength that depends on the average energy levels, but in quantum dots, the discrete energy levels can be changed by the addition or subtraction of just a few electrons—in the crowded conditions of a bulky material, one electron more or less would not be noticed, but this is not true in quantum dots. By varying the size, researchers can easily make quantum dot LEDs emit a variety of wavelengths.

National Nanotechnology Initiative

Sorting out priorities among the many branches of science and technology that are competing for attention and funding is not an easy task. The United States established the National Science and Technology Council (NSTC) in 1993 to advise the president on goals and strategic pursuits in the various fields. Soon thereafter, the potential and promise of nanotechnology brought attention to this field of research, and in the 2001 federal budget, nanoscale science and technology came into the spotlight when it was identified as being an initiative—the National Nanotechnology Initiative—which means the government is firmly committed to funding this research. The Nanoscale Science, Engineering and Technology Subcommittee, a component of NSTC, was established to guide policy and make recommendations.

One of the most important jobs of the National Nanotechnology Initiative is to coordinate the efforts of the large number of government agencies that fund scientific and technological research. Nanotechnology involves not just chemistry but also physics, biology, engineering, materials science, and other disciplines, so collaboration is critical. Twenty-six federal agencies participate in the initiative, 13 of them with significant research budgets. In 2007, for example, these 13

As these experiments demonstrate, discrete levels and quantum mechanics strongly affect the nature of quantum dots. Although quantum dots are composed of hundreds of atoms, the dots are so small that they exhibit quantum properties normally observed only on the atomic level. Researchers who pack together nanocrystals into slightly larger solids have found that the quantum effects can be maintained. The University of Pennsylvania scientists Hugo E. Romero and Marija Drndic reported in 2005 that lattices of nanocrystals can make tiny solids with a variety of properties, some of which are adjustable by varying the size, shape, and composition of the crystals. The researchers crafted quantum dots

agencies contributed an estimated $1.35 billion in support of nanotechnology research; the bulk of the money came from the Department of Defense, National Science Foundation, Department of Energy, and National Institutes of Health, but the National Aeronautics and Space Administration and the National Institute of Standards and Technology also heavily invested in this field of research. The diversity of these agencies testify to the broad potential applications of nanotechnology.

Five major themes underlie the National Nanotechnology Initiative:

- fundamental nanoscale research
- "grand challenges," such as nanomaterial development, that are deemed priorities
- facilities and equipment—the infrastructure of research
- ethical, legal, and social implications of nanotechnology
- establishment of centers of excellence

Some of the funding for facilities, centers, and experiments has flowed to universities, while other funding has gone to government laboratories. For example, the Center for Functional Nanomaterials was established at Brookhaven National Laboratory in New York. Work done in this laboratory by Mathew Maye and Oleg Gang has been discussed earlier in this chapter.

made of lead (Pb) and selenium (Se), and they adjusted the packing to turn the solid from a nonconductor—in which coulombic (electrical) interactions are blocked—into a semiconductor—in which conduction electrons "hop" among energy levels. The paper entitled "Coulomb Blockade and Hopping Conduction in PbSe Quantum Dots" appeared in *Physical Review Letters*.

Nanomaterial research is one of the richest branches of nanotechnology. But scientists and government officials have been impressed with the potential of all the approaches to small-scale science and technology. Achieving this potential will require a lot more research—and money to fund this research. To help meet this need, the U.S. government launched the National Nanotechnology Initiative in 2001, as described in the sidebar on pages 56–57.

One of the most important applications of nanotechnology is medicine. Biology involves many small-scale objects such as the fundamental unit of life—the cell—most of which are about 0.001–0.004 inches (0.0025–0.01 cm) in diameter, and proteins, the large molecules that perform a variety of important functions in the body. Many of the diseases that afflict people arise because of problems with certain cells or proteins.

NANOTECHNOLOGY IN MEDICINE— CANCER DETECTION AND TREATMENT

Although nanotechnology has a large number of applications in medicine, this section will focus on cancer—the result of a cell or small group of cells that grows out of control.

Growth is essential during development, as a baby matures into an adult, and is also required in the maintenance and repair of an adult's tissues. Cells make new tissue by growing and dividing—a cell divides into two "daughter" cells, which generally have the same properties as the original cell. Division continues until the needed tissue is complete.

Cellular division is a highly regulated process, subject to checks and balances to ensure that growth occurs at a timely pace and in the correct amount. Too much tissue is just as bad as too little, and an out-of-control cell could quickly divide and consume all of the body's nutrients and available space. To prevent this, certain proteins monitor and regulate division. If something goes wrong and a cell begins to evade these controls and divide needlessly, a special set of genes in the cell will be activated to kill it. Self-destruction is a drastic measure, but the suicide of an individual cell is better than the death of the whole body.

A problem occurs when a cell not only manages to escape the normal controls but also refuses to take a bullet, so to speak, for the team. This could happen when a series of mutations—unintentional alterations in the cell's genes—interferes with the checks and balances. Since cells pass on their genes during division, the new cells will also possess these mutations, along with the ability to escape regulatory control. The result is a tumor, which is a mass of unwanted cells. Sometimes the growth is constrained, in which case it is a benign tumor that often poses little threat and can usually be removed. But in some cases the cells spread and invade other tissues in the body. This is the hallmark of cancer, the second-leading cause of death in the United States (after heart disease). Cancer-causing agents such as ultraviolet light, nuclear radiation, and tobacco products cause gene mutations that lead to cancer.

Once physicians detect cancer in a patient, treatments focus on killing the abnormally dividing cells. These treatments are difficult because anything that kills a cancerous cell also tends to be unhealthy for other cells in the body. If the cancer is caught in the early stages, when there are few cancerous cells and they are localized in a relatively small area, treatment has a good chance of success. Most of the treatment procedures involve a strong dose of radiation or chemicals, which targets rapidly dividing cells. The treatments focus on areas of the body containing a lot of cancerous cells, but some collateral damage, generally called a side effect, often occurs. Healthy cells die along with the cancerous ones, and regions that normally experience a lot of cell division, such as the digestive tract and hair follicles, are most susceptible. Common side effects of cancer treatments include nausea and hair loss.

More precise targeting of cancerous cells would reduce or even eliminate the side effects. Such precision would also allow treatments to use higher doses, becoming much more effective without harming the patient. Targeting specific cells, tissues, or organs is a common goal in medicine—medical treatments and drug dosages for most illnesses have this aim. But side effects abound in almost all cases because the treatment procedures or drug delivery methods are general rather than specific. Cancer-fighting doctors aim for cancerous cells, but healthy cells surround the target and get a blast of "friendly fire." Drugs usually work by entering the circulatory system—the patient receives an injection or swallows a pill in which the medicine eventually gets into the bloodstream—and the blood carries it not only to the site of disease but also to other tissues and organs as well.

Nanotechnology offers the promise of specific targeting in medicine because its size is on the same scale as cells and proteins it needs to find. Instead of flooding the body with medication, medical applications of nanotechnology zooms in on the problem areas. Nanotechnology seeks to hunt down the bad guys while leaving the peaceful population alone.

Cancer is a particularly important disease to treat early, before the disease invades too much tissue. Quantum dots may offer one important method of detecting the disease in its early stages. Cancerous cells often make certain proteins or other molecules that are attached to the cell's membrane and help the cell to grow and divide. These proteins are not found in healthy cells or are much rarer than in cancerous ones and, therefore, can serve as miniature markers for the diseased cells. These markers would make cancer cells easy to detect, but physicians do not usually have an effective tool with which to see them.

Visualizing the markers is a good mission for nanotechnology since quantum dots can be attached to molecules that bind to the markers. Shuming Nie, a researcher at Georgia Institute of Technology, and other scientists have developed quantum dots made of cadmium selenide for this purpose. These quantum dots fluoresce—they emit light when excited by electromagnetic radiation from the ultraviolet portion of the spectrum, which is invisible radiation slightly beyond the violet end—and the color of light depends on the size of the quantum dot. Researchers can make a colorful array of dots by varying the size and structure, then attach each one to a different molecule. When placed in tissue samples, these dots bind to cancer cells. By shining a small, safe amount of ultraviolet light on the sample, the dots—and the cancer cells—stand out in full color.

Physicians are perfecting this technology in order to detect cancers at the soonest possible moment. But nanotechnology may not only be able to find cancer cells, it may also provide the means to attack and destroy them.

Gold offers yet another example of a case where a small-scale material has properties not shared by its bulkier form. Gold is an inert metal (though a shiny and valuable one), but it makes quite a boisterous

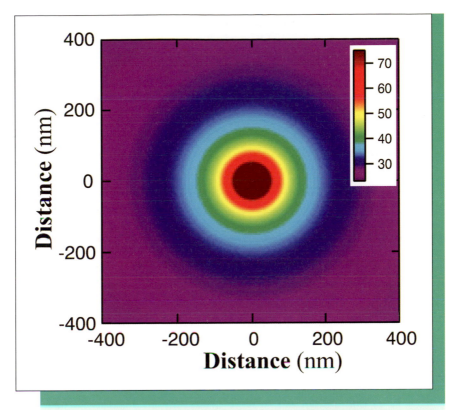

This color-coded image shows a gold nanoparticle (dark red) heated with an infrared laser beam. The surroundings get hot as well, but the temperature fades quickly to 81°F (27°C), represented by the color purple. (The color scale is given in Celsius.) *(Yeonee Seol/JILA)*

nanoparticle. Nanoparticles of gold can be connected to the same molecules that bind to the cancer markers. When attached to the cancerous cells, gold nanoparticles can provide the means by which physicians selectively destroy the right cells. Mostafa A. El-Sayed, another researcher at Georgia Institute of Technology, found that aiming a laser beam at tissues causes the death of cells with attached gold nanoparticles. The intensity of the laser is low enough that cells without the nanoparticles are unharmed.

Other scientists have also been experimenting with this effect. Jennifer L. West and her colleagues at Rice University designed small shells, "nanoshells," of gold. The nanoshells have a core made of a dielectric material—which does not conduct electricity but affects electric fields—surrounded by a shell of metal. By varying the structure and composition of the shell, the researchers can "tune" the nanoparticle, making it resonate at certain frequencies, which makes the nanoparticle more responsive to these frequencies. Some frequencies, such as those that lie in the infrared portion of the spectrum—an invisible band of frequencies next to red—pass through the body because they are not readily absorbed by cells and tissues. When nanoshells are resonant at these frequencies, imaging them with visual equipment becomes more effective. The increase in effectiveness also occurs when researchers zap the particles with a laser, a process called photothermal because of the destruction from the heat (thermal energy) that was induced by the laser light (photons).

Ji-Xin Cheng, a scientist at Purdue University in Lafayette, Indiana, is also interested in this technology. He has used tiny rods of gold, called nanorods, to tag and destroy cancerous cells. Yet Cheng has discovered that heat, although important, only initiates the process that eventually leads to the cell's death. In a 2007 paper, "Gold Nanorods Mediate Tumor Cell Death by Compromising Membrane Integrity," published in *Advanced Materials,* Ling Tong, Cheng, and their colleagues report an experiment in which they show that the nanorods get hot when illuminated, as expected, but this heat causes the formation of tiny bubbles of hot gas. This process, known as cavitation, results in little explosions that puncture the cell's membrane. The integrity of a cell's membrane is critical to its survival. When holes or gaps appear, the contents of the cell mingles with the extracellular fluid, a substance that has a much different character than the fluid inside the cell. Nutrients escape, damaging molecules pour in, and the cell dies.

Nanotechnology's small scale is vital to its medical applications, as it is in all other applications. Defenses of the body's immune system recognize and attack foreign material. This system is important in fighting off infection, but the immune system did not evolve to distinguish the efforts of physicians. If a treatment requires the introduction of large objects that must stay in the body for a while, such as transplanting an organ from a donor into the patient's body, the treatment may fail because of an immune system attack. But nanotechnology may be successful where other techniques fail, because tiny particles in the nanoscale range slip by the immune system. This is yet another reason why the miniaturization of technology has tangible advantages.

Like most of the techniques described in this chapter, these medical procedures are still in the experimental stage. New medical treatments always require extensive testing before physicians widely adopt them, and a considerable amount of time may elapse before these techniques move from the laboratory to the hospital. But the potential for tremendous advances in medicine is clear.

CONCLUSION

Much of chemistry occurs at small scales. As Dalton and other scientists realized, atoms are the basic units of matter and combine chemically to form molecules. Although chemists usually work with substantially larger amounts of matter, researchers in the field of nanotechnology are beginning to develop the techniques to manipulate matter on the atomic and molecular scale. By finding the right conditions in which atoms and molecules will assemble into functional structures, or by constructing tiny machines that oscillate, researchers have entered the domain of the atom.

Nanomaterials for medical applications have been found, but researchers' ambitions extend well beyond this. Most diseases are detected only after they have gained a foothold, when the patient begins to display symptoms. Physicians are consequently forced to play catch-up. As is true in cancer, treatments are generally much more effective in the early stages of a disease, before the pathology—the disease—has had time to progress and spread. Better yet, preventing the disease from ever getting started avoids the expense of a treatment, as well as any side effects it may cause.

Doctors cannot be on guard at every moment, but a nanobot might. Imagine a squadron of tiny machines as they patrol the body. The nanobots swim through the bloodstream and the extracellular spaces, inspecting cells and proteins in search of any signs of damage or pathological situations or agents. When they spot their quarry, they automatically attack, perhaps replicating themselves if reinforcements are needed.

As of now, building such nanobots is not within nanotechnology's capacity. But there is no reason why these machines could not one day be developed, as Richard Feynman pointed out in his 1959 lecture. The mechanisms of diagnosing or detecting diseases on the molecular scale is advancing rapidly, and researchers are finding ingenious ways of miniature propulsion. Certain bacteria, for instance, propel themselves through water by swinging a whip, known as a flagellum, which has inspired researchers such as Bahareh Behkam and Metin Sitti in the NanoRobotics Laboratory at Carnegie Mellon University in Pittsburgh, Pennsylvania. These scientists have attached 0.0004-inch (0.001-cm) plastic beads to flagellating bacteria. The bacteria haul the beads like horses pulling a chariot, controlled by the addition of chemicals to turn the flagellum on or off. The research was reported in "Bacterial Flagella-Based Propulsion and On/Off Motion Control of Microscale Objects," published in *Applied Physics Letters* in 2007.

The notion of miniature robots swooping down and obliterating a misbehaving cell may seem fantastic—rather like the science-fiction film *Fantastic Voyage,* mentioned earlier in the chapter—but it is a feasible goal. Nanotechnology has received a lot of publicity, including portrayals in movies that have fired the imagination of scientists and engineers, as well as being the subject of many books—the list in the Further Resources section of this chapter is just a sample of what has been published. This field of research may find it difficult to live up to such high expectations. But nanotechnology is a frontier of science that has been making impressive strides, and it promises much more to come.

CHRONOLOGY

ca. 400 B.C.E.	Greek philosopher Democritus (ca. 460–370 B.C.E.) proposes the theory that matter consists of tiny, indivisible atoms.
1670s	Dutch craftsman and scientist Antoni van Leeuwenhoek (1632–1723) manufactures high-quality microscopes and discovers bacteria and other small-scale creatures and objects.
1803 C.E.	British chemist John Dalton (1766–1844) introduces the modern concept of atoms as one of the principles of chemistry and physics.
1857	British scientist Michael Faraday (1791–1867) experiments with colloidal gold—a suspension of extremely small gold particles in a liquid. These

experiments sparked interest in small particles and their properties.

1897   British physicist Sir Joseph John Thomson (1856–1940) discovers the electron.

1920s   German physicist Werner Heisenberg (1901–76), Danish physicist Niels Bohr (1885–1962), Austrian physicist Erwin Schrödinger (1887–1961), building on the work of German physicists Max Planck (1858–1947) and Albert Einstein (1879–1955), discover the principles of quantum mechanics.

1930s   German engineers Enrst Ruska (1906–88) and Max Knoll (1897–1969) develop the first electron microscope.

1958   Texas Instruments researcher Jack Kilby (1923–2005) invents a tiny electronic circuit known as an integrated circuit, which rapidly becomes an important component of computers and other electronic devices.

1959   American physicist Richard Feynman (1918–88) delivers a lecture, "There Is Plenty of Room at the Bottom," advocating nanoscale science.

1974   Japanese physicist Norio Taniguchi coins the term *nanotechnology*.

1980s   American Louis Brus, a researcher at Bell Laboratories, and Russian scientists Alexander Efros and Aleksey Ekimov discover and investigate quantum dots.

1981   IBM researchers Gerd Binnig and Heinrich Rohrer develop the first scanning tunneling microscope.

1986   Eric Drexler writes a book, *Engines of Creation*, which draws much attention to nanotechnology.

1990s	Nanomaterials begin appearing in sunscreens and other products.
2001	The U.S. government establishes the National Nanotechnology Initiative.
2003	University of California at Berkeley physicist Alex Zettl and his colleagues build the first nanoscale electronic motor.
2007	Daniel S. Johnson, Michelle Wang, and their colleagues measure the force exerted by a helicase molecule.

FURTHER RESOURCES
Print and Internet

Alper, Joe. "Self-Assembly: Getting Molecules to Put Themselves Together to Make Nanoscale Devices." Available online. URL: http://nano.cancer.gov/news_center/monthly_feature_2005_jul.asp. Accessed May 28, 2009. This feature article, published in a newsletter for the National Cancer Institute, offers an accessible tutorial on self-assembly and supramolecular chemistry.

Benhkam, Bahareh, and Metin Sitti. "Bacterial Flagella-Based Propulsion and On/Off Motion Control of Microscale Objects." Available online. URL: http://scitation.aip.org/vsearch/servlet/VerityServlet?KEY=APPLAB&smode=results&maxdisp=10&possible1=bacterial+flagella-based&possible1zone=article&OUTLOG=NO&viewabs=APPLAB&key=DISPLAY&doc ID=1&page=0&chapter=0. Benhkam and Sitti attached microscopic plastic beads to bacteria, which haul the beads under control of certain chemicals. Accessed September 24, 2009.

Berube, David M. *Nano-Hype: The Truth Behind the Nanotechnology Buzz.* Amherst, N.Y.: Prometheus Books, 2005. People get excited about new ideas and innovations—sometimes too excited. Although enthusiasm helps push the boundaries of science, it can also raise false hopes and unrealistic expectations, as well as ring alarm bells over threats and dangers that are highly improbable. Nanotechnology is no exception. Berube's book examines this field of research from a critical viewpoint.

Drexler, K. Eric. *Engines of Creation.* New York: Bantam Doubleday Dell, 1986. Drexler's book fired the imagination of researchers as it promoted the new subject of nanotechnology. The book offers insights into the early stages of nanotechnology research, though it can be challenging to read at times.

Feynman, Richard. "There Is Plenty of Room at the Bottom." Available online. URL: http://www.zyvex.com/nanotech/feynman.html. Accessed May 28, 2009. This Web page contains the text of Feynman's 1959 pioneering lecture on small-scale science. The lecture is creative and entertaining, typical for the late and much-missed Nobel laureate, who passed away on February 15, 1988.

Johnson, Daniel S., Lu Bai, Benjamin Y. Smith, Smita S. Patel, and Michelle D. Wang. "Single-Molecule Studies Reveal Dynamics of DNA Unwinding by the Ring-Shaped T7 Helicase." *Cell* 129 (2007): 1,299–1,309. By using a laser beam, the experimenters made precise measurements of the movement of the bead, observing the forces imposed by helicase enzymes.

Maye, Mathew M., Dmytro Nykypanchuk, Daniel van der Lelie, and Oleg Gang. "A Simple Method for Kinetic Control of DNA-Induced Nanoparticle Assembly." *Journal of the American Chemical Society* 128 (2006): 14,020–14,021. The researchers created nanoparticles by attaching the components to complementary strands of DNA.

National Institute of Standards and Technology. "Nanotechnology is BIG at NIST." Available online. URL: http://www.nist.gov/public_affairs/nanotech.htm. Accessed May 28, 2009. NIST is a federal agency devoted to aiding innovation and industrial output by advancing the science of measurement and providing accurate standards by which such measurements can be made. Work at the NIST laboratory involves atomic and molecular devices and other aspects of nanotechnology, as described on this Web site.

Regan, B. C., S. Aloni, K. Jensen, and A. Zettl. "Surface-Tension-Driven Nanoelectromechanical Relaxation Oscillator." Available online. URL: http://scitation.aip.org/vsearch/servlet/VerityServlet?KEY=APPLAB &smode=results&maxdisp=10&possible1=surface-tension-driven%5

D&possible1zone=article&OUTLOG=NO&viewabs=APPLAB&key=DISPLAY&docID=1&page=0&chapter=0. Accessed May 28, 2009. The researchers report the development of a tiny oscillating device.

Romero, H. E., and M. Drndic. "Coulomb Blockade and Hopping Conduction in PbSe Quantum Dots." Available online. URL: http://prola.aps.org/abstract/PRL/v95/i15/e156801. Accessed September 24, 2009. The researchers crafted quantum dots made of lead and selenium, and adjusted the packing to turn the solid from a nonconductor into a semiconductor.

Sargent, Ted. *The Dance of Molecules: How Nanotechnology Is Changing Our Lives.* New York: Thunder's Mouth Press, 2005. This highly optimistic book, written by a researcher, lucidly explains the science of nanotechnology as well as its practitioners and its many applications.

Scientific American. Understanding Nanotechnology. New York: Warner Books, 2002. Containing a special set of essays originally published in *Scientific American,* this book explores molecular motors, nanotubes, medical applications, and much more.

Tong, Ling, Yan Zhao, Terry B. Huff, Matthew N. Hansen, Alexander Wei, and Ji-Xin Cheng. "Gold Nanorods Mediate Tumor Cell Death by Compromising Membrane Integrity." *Advanced Materials* 19 (2007): 3,136–3,141. The researchers show that nanorods get hot when illuminated, as expected, but this heat causes the formation of tiny bubbles of hot gas.

Web Sites

Carnegie Mellon University: NanoRobotics Lab. Available online. URL: http://nanolab.me.cmu.edu/. Accessed May 28, 2009. The NanoRobotics Lab at Carnegie Mellon University in Pittsburgh, Pennsylvania, is exploring wall-climbing robots, water-running robots, water-striding robots, swimming robots, and other tiny mechanical machines, as outlined on this Web site.

Center for Responsible Nanotechnology. Available online. URL: http://crnano.org/. Accessed May 28, 2009. This nonprofit group aims to explore the rewards as well as the risks of nanotechnology, raising public awareness of the issues and informing government policies and funding strategies. The group's Web site includes discussions of the problems as well as possible solutions.

Foresight Nanotech Institute. Available online. URL: http://www.foresight.org/. Accessed May 28, 2009. Founded in 1986 and based in Menlo Park, California, this nonprofit organization aims to promote the beneficial aspects of nanotechnology and inform the public on the achievements and the realistic goals in this field of research. Foresight Nanotech Institute's Web site contains many articles with current news and basic scientific information on nanotechnology, as well as a roadmap for future developments.

IEEE Virtual Museum. Available online. URL: http://www.ieee-virtual-museum.org/. Accessed May 28, 2009. The Institute of Electrical and Electronics Engineers (IEEE) maintains this Web site to provide "virtual" (Web-based) exhibits of a variety of electrical and electronic topics. Although all of the exhibits are interesting, the two that are most relevant to this chapter are "Let's Get Small: The Shrinking World of Microelectronics," and "Small is Big: The Coming Nanotechnology Revolution."

NanoKids. Available online. URL: http://cohesion.rice.edu/naturalsciences/nanokids/index.cfm. Accessed May 28, 2009. NanoKids is an educational program headed by James M. Tour, a chemistry professor at Rice University. Geared for young students, the Web site illustrates the principles of nanoscience with clever animations and witty explanations.

National Nanotechnology Initiative. Available online. URL: http://www.nano.gov/. Accessed May 28, 2009. The Web site for the National Nanotechnology Initiative, formed in 2001 to coordinate federal government funding for nanotechnology research, presents facts and figures on nanotechnology, updated research news, discussions of societal and safety issues, information on research centers, and much more.

CHEMISTRY OF THE BRAIN: MOLECULES, MOOD, AND MENTAL ILLNESS

One of the first people to apply science to medicine was the ancient Greek physician Hippocrates (ca. 460–377 B.C.E.). Influenced by the idea that the world is composed of four substances—earth, air, fire, and water—as taught by the Greek philosopher Empedocles (ca. 495–435 B.C.E.), Hippocrates proposed that four fluids are critical in determining a person's state of health. These fluids, known as humors (from a Latin term for *moisture*), were called blood, yellow bile, black bile, and phlegm. According to Hippocrates, an imbalance in these humors caused disease. Later, people associated a specific temperament or personality with these humors, a theory that was one of the earliest attempts to explain moods and emotions. Blood, for example, was associated with an optimistic disposition, while black bile corresponded to depression.

But eventually people realized that the brain rather than fluids governs mood and temperament—an injury to the brain can lead to serious changes in personality. In 1848, for example, an accidental explosion in Vermont sent an iron rod into the front part of the skull of Phineas Gage (1823–60), a 25-year-old railway worker. Gage miraculously survived and recovered with little apparent injury. Yet, in temperament he was not the same man; before the accident he had been polite and reliable, whereas afterward he was careless and offensive.

Although scientists have studied the brain for many years, no one knows exactly how it works. Some people are generally happy, some people tend to be gloomy, and some are susceptible to serious problems such as depression or drug addiction. These differences in personality reflect some kind of difference in the brain. While the exact nature remains a mystery, researchers who study the chemistry of the brain have uncovered interesting clues. These clues led to the study of signals and messages occurring in the brain, and the chemicals involved in transmitting them. This chapter focuses on research on these transmissions and the role they play in certain disorders such as depression, schizophrenia, and drug abuse.

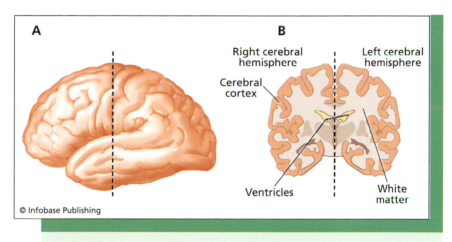

© Infobase Publishing

On the left (A) is a profile of the human brain, and on the right, a magnified view (B) of a slice through the brain as indicated by the dotted line. The anatomy of the brain consists of two hemispheres, each covered with multiple layers of important cells known as the cerebral cortex. White matter, which consists of "wiring"—projections from one cell to another—lies beneath the cerebral cortex, along with some groups of cells that perform various functions. The ventricles are cavities through which runs a substance called cerebrospinal fluid, which helps protect and cushion the delicate structures of the nervous system.

INTRODUCTION

The figure on page 71 shows a diagram of an adult human brain, which weighs about three pounds (1.4 kg) on average. Early chemists who studied the brain did not have instruments or techniques to separate the components. All the chemists could do was crudely separate some of the material by repeatedly using a *solvent*—a substance that dissolves certain compounds—and other agents that bind to compounds, making them insoluble.

Beginning in the 18th century and continuing through the 19th century, the German scientist Johann Hensing (1683–1726), the French scientists Michel-Augustin Thouret (1748–1810) and Louis-Nicolas Vauquelin (1763–1829), and others discovered that the brain contains a lot of fatty components. Fats are large molecules made of mostly carbon and hydrogen, and they generally feel oily or buttery to the touch. About half of the dry weight of the brain is composed of fats and lipids (fatty compounds). (Dry weight means that the water has evaporated or been removed.) Much of this fatty material was found in the brain's "white" matter, which was named because it is generally pale, particularly when the brain is removed and hardened with chemicals for study. (Brains are so soft they cannot even support their own weight—one of the reasons why brains must be encased in skulls for protection and surrounded by fluid contained within cavities called ventricles and a porous structure known as the meninges.) Researchers later showed that white matter consists of the projections by which brain cells communicate with one another. The fatty molecules wrap around these projections, acting as a kind of insulation and aiding the communication process.

The brain's "gray" matter contains mostly cells. A cell is the fundamental unit of life in all organisms and contains enzymes, nutrients, and genetic material within a thin membrane made of lipids and protein. The human body contains cells that have different properties to perform different functions, such as skin cells, brain cells, muscle cells, and so on. Brain cells are not actually gray but are mostly transparent, as are most cells; however, collections of cells may appear gray or grayish brown under certain circumstances, especially after the brain is chemically treated. Two main types of cell exist in the brain—a glial cell serves important maintenance and support functions, and a *neuron* processes information, possibly aided by glial cells. The human brain contains about a trillion neurons and several times that number of glial cells.

Difficulties in the chemical studies of brain tissue plagued researchers for many years. By examining the brains of recently deceased humans, the French researcher Jean-Pierre Courbe reported in 1833 that normal brains consisted of 2–2.5 percent of the element phosphorous, whereas the brains of people with low intelligence had a lower percentage, and the brains of the "insane" had a higher one. If true, this finding suggested that phosphorous had an excitatory effect on the nervous system—too little of the element dampened activity, while too much caused an unnatural outburst. But the finding was not true, evidently because of mistakes in the difficult procedures of isolating and chemically analyzing tissue. Although mistaken, Courbe's research was an interesting prelude to a similar set of findings in the 20th century, described below.

A great advance in brain science occurred in 1873, when the Italian physician Camillo Golgi (1843–1926) developed a method of staining tissue with silver nitrate. Neurons, as shown in the figure on page 74, have long, thin projections called axons, which often course through and compose the brain's white matter. A typical neuron has a diameter of about 0.002 inches (0.005 cm), though it generally covers a larger area with its branches called dendrites, which receive input from other neurons. The axon projects to other neurons or muscles that may be nearby or, in extreme cases, as far away as 40 inches (100 cm) in humans (and longer in bigger animals). Prior to Golgi's staining technique, biologists could not make a lot of progress studying brain tissue with microscopes because the transparent cells and projections were impossible to track. But the stain filled the whole cell, including the axon, allowing scientists to pick it out of the drab background. (For some as yet unknown reason, applying Golgi's technique to brain tissue stains only a fraction of neurons. This selectivity is essential, otherwise the stain would cover everything and biologists would be unable to pick out a cell in the uniformly dark background.)

With tools such as Golgi staining, researchers studied neurons and their projections. But it was not obvious if the projections made contact with the target neuron or stopped just before reaching the neuron, leaving a small gap. Some researchers, including Golgi, believed that the brain was a giant net consisting of cells in physical contact with one another; other researchers, such as the Spanish anatomist Santiago Ramón y Cajal (1852–1934), were convinced that a gap existed between axons and their target neurons. Cajal realized the brain might be simpler to understand if the cells were joined as one, as Golgi believed, but Cajal was certain the gaps existed. In a lecture Cajal gave when presented with the 1906 Nobel Prize (which he shared with Golgi), Cajal said, "It would be very convenient and very economical from the point of view of analytical effort if all the nerve centres were made up of a continuous intermediary network between the motor nerves and the sensitive and sensory nerves. Unfortunately, nature seems unaware of our intellectual need for convenience and unity, and very often takes delight in complication and diver-

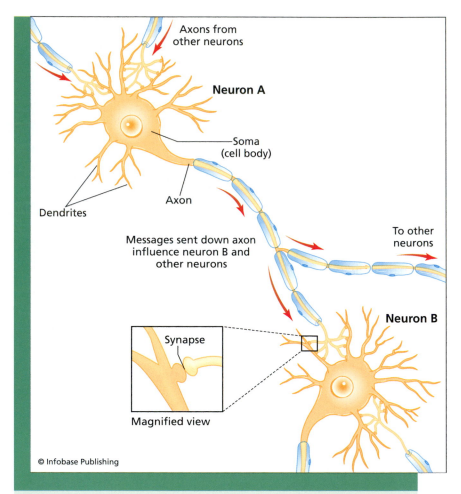

Neuron A, shown at left, receives synaptic inputs from many neurons and sends its axon to make synapses with neuron B and others. Extensions called dendrites branch out from the cell body—the soma—and are the sites of many synapses. (One synapse is shown magnified.)

sity." Improvements in microscopes and staining techniques eventually resolved the debate in favor of Ramón y Cajal by the early 20th century.

Meanwhile, scientists studied the electrical activity of neurons. They discovered that the messages neurons pass to one another consist of brief electrical impulses known as *action potentials*. Neurons (and most other cells) have a small electrical potential of about 60 millivolts between their interior and exterior—this voltage, which is only about 4 percent as strong as a typical flashlight battery, exists across the cell's membrane. The watery solution inside a cell is different from the solution outside the cell, but both solutions contain various nutrients as well as charged particles called *ions*. Ions come from compounds that dissolve in water, such as sodium chloride (table salt) that separates into a positively charged sodium ion and a negatively charged chloride ion. The reason salt dissolves is because the chemical bond between sodium and chlorine is ionic—the bond forms when sodium donates an electron to chlorine. Water molecules pull apart the bond, and the electron stays with the chloride ion, giving it an excess negative charge while leaving sodium with a deficit.

What makes a neuron special is the presence of protein molecules that are sensitive to the voltage across its membrane. Membranes do not allow ions to pass—one of the main jobs of a membrane is to regulate the flow of substances into and out of the cell—but certain proteins embedded in the membrane contain channels to allow ions to pass through. The flow of ions constitutes an electrical current. In neurons, these proteins, known as ion channels, can open and close quickly, producing currents that briefly change the electrical potential of the cell. This process gives rise to an action potential, which lasts only a few milliseconds and travels down the neuron's axon.

But how does the action potential bridge the gap between the axon and the target neuron? The answer to this question lies at the foundation of many important topics in brain chemistry.

SYNAPTIC TRANSMISSION

Action potentials generated by neurons convey information in their rate—how many occur in a given period of time, which can be up to a few hundred per second—and sometimes in their timing relative to one another. Networks of neurons process sensory information, control movement, and create the still mysterious nature of consciousness by sending messages composed of action potentials. As scientists began to study this process, they realized that the transmission of messages across the gap was extremely important. The British physiologist Sir Charles Sherrington (1857–1952) coined the term *synapse* in 1897 to describe this gap. In most cases, the gap is less than 0.000001 inches (0.0000025 cm).

Early researchers thought of two means by which synaptic transmission could occur. Action potentials could jump the gap by some electrical mechanism such as induction—this is the same process by which an electrical current in one coil of a transformer induces a current in the other coil, even though the coils are not in contact. Another possible mechanism of synaptic transmission involves the use of a chemical intermediate. Suppose an action potential caused molecules of a certain chemical to be released at the tip of the axon. These molecules would diffuse across the gap, reaching the other side in a short period of time. But researchers were unsure what effect these molecules could have on the postsynaptic neuron that receives the message. (The transmitting neuron is known as the presynaptic neuron.) In 1906, the British researchers Thomas Elliot (1877–1961) and John Langley (1852–1925) considered the idea of a *receptor* on the postsynaptic neuron, on which the chemical might dock and exert its effects.

This idea of chemical transmission in the nervous system was not widely accepted until researchers began conducting ingenious experiments. In 1921, Otto Loewi (1873–1961), a German-born physician who later immigrated to Austria and then to the United States, experimented with a nerve known as the vagus nerve. A nerve consists of a bundle of axons—these axons constitute the projections of a group of neurons to another group of neurons or to muscles (such as heart muscle or skeletal muscle) or glands (such as the adrenal gland, which sits above the kidney and releases hormones into the bloodstream). The vagus nerve contains projections from neurons in the brain to the heart and is the means by which the brain can regulate heart function (for instance, speeding the rate in times of excitement). As in the case of neuron-to-neuron transmission, messages conveyed to the muscles and glands must cross synapses.

Loewi used frogs in his experiments. He exposed the vagus nerve and heart of one frog, then stimulated the nerve artificially by shocking it with an electrical current. The heart began to beat faster. Loewi extracted some of the fluid surrounding this heart and poured it on another heart, in which the vagus nerve had been cut away. The second heart began to beat faster, as had the first, suggesting that some chemical substance was in the fluid and was affecting the heart rate. The experiment indicated that the vagus nerve conveys messages from the brain to the heart by releasing a chemical or chemicals.

Further research attempted to identify these chemicals. In 1929, the British physiologist Sir Henry Dale (1875–1968) summarized the research on chemical transmission and, based on his findings as well as those of his colleagues, argued that a chemical known as acetylcholine was involved. The German physiologist and physician Wilhelm Feldberg (1900–93) and the British researcher Sir John Gaddum (1900–65) showed in 1934 that certain chemicals such as acetylcholine are released in the synapses of nerves, proving that chemical transmission takes place.

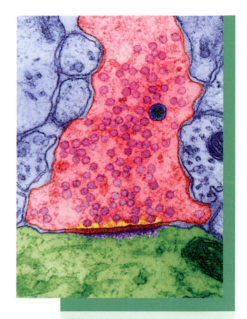

This colored electron microscope image depicts a synapse between the presynaptic embrane, shown in pink, and the postsynaptic membrane, shown in green. The slightly thickened area between the pre- and post-synaptic cells is part of the synapse. Small circles in the presynaptic cell are vesicles containing neurotransmitters. *(Dr. Dennis Kunkel/ Visuals Unlimited)*

Information processing in the brain, therefore, includes electrochemical processes. Action potentials are electrical impulses, but the messages they carry are generally mediated by chemicals. These chemicals are known as *neurotransmitters* since they transmit messages between neurons as well as between neurons and muscles or glands. The following sidebar discusses neurotransmitters in more detail.

The experiments of Loewi, Dale, Feldberg, Gaddum, and their colleagues were conclusive, yet they did not rule out the possibility of other

Neurotransmitters

The list of identified neurotransmitters contains more than 50 substances, although only a dozen or so are common in the brain. To be included, a chemical must be found in the presynaptic neuron and have a definite effect on the postsynaptic target. Neurotransmitter molecules are usually packaged in a container known as a synaptic vesicle, as shown in the figure at right. The vesicle is bound by a membrane, similar to the cell's membrane. Electrical activity initiated by an action potential causes a flow of calcium ions, which in turn cause the vesicle to bind and join the cell's membrane, releasing the contents into the synaptic gap. In the most common situation, neurotransmitters bind to a specific protein embedded in the postsynaptic cell's membrane. This protein is called a receptor for the specific neurotransmitter or ligand it binds. (*Ligand* is a general term for a chemical that binds to a specific target—the word comes from a Latin term, *ligare,* to bind.) The binding of a ligand activates the receptor, which has some sort of effect on the postsynaptic cell—some receptors impact ion channels, and others generate molecules known as second messengers, which in turn activate certain enzymes or other proteins in the cell.

Neurotransmitters include the following substances:

- amino acids: glutamate, gamma-aminobutyric acid (GABA), and glycine
- small molecules, including acetylcholine, dopamine, norepinephrine (also known as noradrenalin), and serotonin (also known as 5-hydroxytryptamine, or 5-HT)
- peptides (short chains of amino acids) such as enkephalin and vasopressin
- soluble gases such as nitric oxide

Many of these substances play additional roles in the body. For example, besides acting as neurotransmitters,

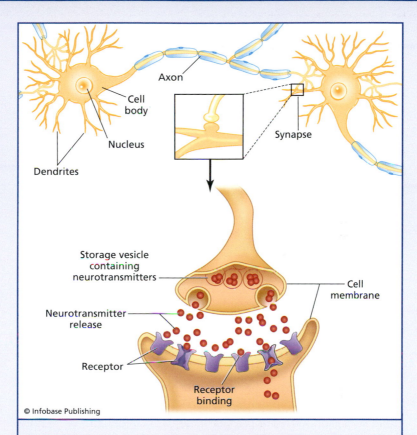

Axon

Cell
body

Nucleus

Dendrites

Synapse

Storage vesicle
containing
neurotransmitters

Cell
membrane

Neurotransmitter
release

Receptor

Receptor
binding

© Infobase Publishing

A magnified view of the synapse, shown in the bottom portion of
the figure, illustrates the release of neurotransmitters when a
vesicle fuses with the cell's membrane and spills the molecules
into the synaptic gap. Diffusing randomly, some neurotransmitter
molecules bind and activate certain receptors embedded in the
recipient neuron's membrane.

amino acids such as glutamate and glycine are also components of proteins.

Synaptic release of neurotransmitter is a quick process
that finishes in a few milliseconds. But before the synapse is

(continues)

(continued)

ready for the next event, most of the released neurotransmitter molecules should be gone, otherwise the postsynaptic effect may linger. In a few cases, this happens, but in most synapses, there is a mechanism that clears the "field" of any residual neurotransmitter. For a few neurotransmitters such as acetylcholine, enzymes lurking in the membranes of the cells break apart the neurotransmitter molecule and prevent further activation of the receptors. Most neurotransmitters, though, are carried back into the presynaptic membrane and recycled for future use. This process is called reuptake.

Students sometimes wonder why there are so many different neurotransmitters. Would only a few suffice instead? The reason nervous systems employ such a bewildering diversity is apparently to obtain a multitude of different effects. Receptors are the means by which the postsynaptic target is affected, and having a wide variety of receptors, activated by different molecules, enables synaptic transmission to accomplish a number of different tasks. Peptides may modulate a neuron's properties and have long-term affects, while receptors for other neurotransmitters are more fast-acting. And, although the figure on page 79 illustrates the common case in which synaptic transmission occurs from a neuron to a specific target, in other cases, the neurotransmitter may have a wider release, affecting a nonspecific group of neurons. This is true with neurotransmitters such as dopamine. One neurotransmitter, nitric oxide, can even act as a retrograde messenger, traveling from the postsynaptic cell to the presynaptic cell.

transmission mechanisms. Although chemical transmission seems to be the most important mechanism, there are synapses in the brain that are mediated not by chemicals but by narrow openings that connect the cells. Transmission across these synapses, which are known as gap junctions, is electrical—the currents flow from one cell to another. Gap junctions play a vital role in certain networks, but they will not be discussed in detail in this chapter.

Straight-through transmission via gap junctions has the advantage of simplicity over the complicated process of chemical transmission. But chemical transmission gives nervous systems a great deal of flexibility. With more links in the chain, brains have more opportunities to make adjustments—there are more "knobs" and "controls" with which to fine-tune the activities of neurons. Current theories of learning and memory identify the plasticity—adjustability—of synapses as the mechanism by which brains acquire and store information. Intelligence evolves from the continual tweaking of chemical transmission in the brain.

But chemical transmission also makes it difficult for researchers to study neurochemistry—the brain's chemistry. Synaptic transmission is brief and occurs within a small portion of the protected confines of the brain. No wonder Thoulet, Vauquelin, Courbe, and other early researchers had such trouble.

MEASURING THE BRAIN'S CHEMICALS

Studying the chemistry and biology of brain function often requires the researcher to find a model organism—an organism that has anatomical or chemical features conducive to study. Loewi, for example, studied chemical transmission in the vagus nerve of frogs. For axons and synapses, many researchers have used the squid, which contains a giant axon with a diameter of up to 0.04 inches (0.1 cm); this may seem small, but it is about 50 times bigger than typical axons. (This particular axon has such a large diameter because its great size helps to speed conduction of the impulse, which in this case carries information to help the animal escape a predator—an action that needs to be done with as much dispatch as the animal can muster.) The associated synapses are also large. The British scientists Sir Alan Hodgkin (1914–98) and Sir Andrew Huxley (1917–) discovered a great deal of information on ion channels in the 1940s and 1950s using this model organism.

Fundamental principles of chemistry also apply to biochemistry—biological chemistry—and the techniques of neurochemistry have improved since the days of Courbe and his phosphorus analysis. As described in chapters 1 and 2, chemical elements combine in chemical reactions to form compounds. In order for chemical reactions to occur, the reactants—the atoms or molecules participating in the reaction—must meet.

Atoms and molecules are always in motion, though the average speed depends on temperature—higher temperatures mean higher average speeds. Under favorable conditions, which usually includes an adequate supply of reactants and a reasonably warm temperature, a reaction may proceed quickly. But in biochemistry, which usually deals with large molecules such as fats and proteins, the reactants not only must meet, they must also be in correct orientation for the reaction to occur. Proteins, for example, are composed of long chains of covalently bonded amino acids and have a three-dimensional shape. A reaction may involve a certain region of the protein, and the other reactants must be in the proper position or the reaction will not take place. Such precise arrangements may seldom happen, even at high temperatures, and if so, the reaction will be rare.

Life and Death of a Neurotransmitter Molecule

Although some neurons release more than one type of neurotransmitter, many neurons make do with only one. The machinery to make, use, and terminate a specific neurotransmitter can be found only in the neurons that employ it (unless the neurotransmitter molecule has other uses in the body). Serotonin, for example, can be found only in a small number of neurons located in the brain stem. (Since serotonin has other functions, it is also found in other places in the body.) But this tiny number of neurons has widespread projections throughout the brain, making serotonin one of the most important chemicals despite the small number of neurons using it as a neurotransmitter.

Neurons manufacture serotonin in a series of chemical reactions, beginning with the amino acid tryptophan. Catalyzing these reactions are specific enzymes, such as an enzyme known as tryptophan hydroxylase, which speeds up a reaction that adds a hydroxyl group (an OH molecule) to tryptophan. Serotonin is packaged into vesicles, with each vesicle

Organisms deal with this situation by speeding up reactions with catalysts called enzymes. A catalyst affects the rate of a reaction but does not otherwise participate, so it is not chemically altered. Enzymes are usually proteins that temporarily bind the reactants in such a way as to bring them together in the correct position. This binding is not done with strong bonds such as covalent or ionic bonds, but with weaker attractions that are more easily broken. An enzyme usually catalyzes only one specific reaction since its shape and composition are generally such that it binds only a specific set of reactants.

Enzymes are also excellent ways for biological organisms to regulate the rate of reactions. An organism's *metabolism* consists of all the various reactions in that organism. Thousands of reactions take place

containing a few thousand serotonin molecules. Special proteins transport the vesicles down axons to the release sites. (For serotonin, these release sites are swellings on the axon known as varicosities.) An action potential causes the vesicle to fuse with the cell's membrane and releases serotonin into the gap. The molecules diffuse away, and only a few reach the other side—the random paths take many of them off target, just as only a few pellets of a shotgun blast may hit the mark. The serotonin molecules that reach the target dock to specific receptors, called serotonin receptors, which are proteins embedded in the membrane. But many of the other serotonin molecules are not wasted—proteins called transporters, embedded in the axon's membrane, bind to serotonin and transport it across the membrane, retrieving some of those that wandered off target.

Neurons guard against accumulating too much serotonin with the use of enzymes such as monoamine oxidase (MAO). This enzyme catalyzes a reaction converting serotonin to a different molecule. Some of these serotonin *metabolites*—compounds derived from serotonin—are excreted by the body, marking the end of the road for these molecules.

as the organism extracts energy from food, maintains its structures and tissues, and repairs any damage. These reactions must be carefully orchestrated so that the products are present when needed, but wasteful overabundances are avoided. Temperature is important—the human body maintains a steady temperature of about 98.2°F (36.8°C)—but organisms can also vary the reaction rates by producing more or less of the necessary enzymes.

As reactions proceed, reactants are chemically combined into compounds, which then perform their functions. These compounds are eventually consumed in other reactions, and their fragments are used to generate new molecules or are excreted. The life cycle of chemicals in the body is a critical aspect of biochemistry, though neurochemistry researchers are hampered by a limited access to the brain. The following sidebar on pages 82–83 describes a neurotransmitter's metabolic journey.

Because of the skull, researchers cannot easily sample the brain's chemicals in most organisms. Another hindrance is the blood-brain barrier, which the body uses to control the substances that enter and exit the brain. Oxygen and glucose are welcome, but many other substances flowing in the blood are not. Small molecules can slip in and out, but others, such as serotonin, are too big. (This is the reason why neurons must manufacture serotonin—the blood contains this large molecule, but it cannot get into the brain.)

The blood-brain barrier foils most efforts to use the blood to measure the brain's chemistry, but researchers can get around this obstacle by sampling cerebrospinal fluid (CSF). CSF is the fluid that circulates in the meninges of the brain and spinal cord and keeps the delicate tissues from getting rattled around and damaged in their hard, bony container. The brain makes CSF from blood, and certain metabolites get mixed in. One of these metabolites is 5-hydroxyindole acetaldehyde (5-HIAA), a major metabolite of serotonin. Researchers who carefully puncture the meninges and extract a sample are rewarded with information concerning serotonin levels in the person's brain, as described below.

With a limited ability to sample the brain's chemicals directly, researchers have turned to other methods. As described in the sidebar on pages 82–83, proteins such as enzymes and receptors play a vital role in neurotransmitter function. Since these proteins are specific to a specific neurotransmitter, they serve as markers for its use. By attaching various molecules such as dyes, radioactive substances—which emit energetic particles or radiation—or fluorescent molecules—which emit light—to these proteins or to molecules that manufacture or bind them, researchers can locate and identify specific neurotransmitters and the neurons that use them. These methods will be described shortly.

Observant researchers can also find clues by noticing the effects of certain drugs. Reserpine is a drug that has been used to treat high blood pressure, among other disorders, but in the 1950s, a few years after it was isolated and marketed, physicians noticed that about 15 percent of the patients who took it became depressed. Subsequent research showed a side effect—an unintended consequence—of the drug was to make certain synaptic vesicles porous. Neurotransmitter molecules leak out, reducing the effectiveness of the transmission. This happened for the neurotransmitter group known as the monoamines—dopamine, norepinephrine, and serotonin. The finding was one of the first links between neurotransmitter function and mood disorders.

NEUROTRANSMITTERS AND DEPRESSION

Diagnosing depression is not a simple matter. Everyone experiences sadness once in a while, and in certain situations, such as the loss of a loved one, it is expected. The American Psychiatric Association, a group of professional psychiatrists, publishes the *Diagnostic and Statistic Manual of Mental Disorders,* a manual that outlines the criteria for diagnosing psychiatric disorders. This widely followed manual is periodically updated, and it is currently in its fourth edition. DSM-IV, as it is often abbreviated, defines a major depressive episode to be when the patient shows a depressed mood or the absence of pleasure for a certain period of time, as well as exhibiting some other symptoms such as loss of sleep, appetite, or recurrent thoughts of suicide. Depression is one of the most common disorders, with millions of cases diagnosed every year in the United States alone.

About the same time as the reserpine finding, physicians noticed that some of the drugs used to treat other diseases appeared to have a beneficial side effect—raising the patient's mood. Upon further testing, a chemically modified version of one of these drugs effectively reduced the symptoms of depressed patients. This drug, iproniazid, inhibits MAO, the enzyme that destroys the monoamine neurotransmitters—dopamine, norepinephrine, and serotonin. As a result, more of these neurotransmitters are available. In 1958, many physicians and psychiatrists began using iproniazid as a treatment for depressive illness.

Reserpine and iproniazid research led to the monoamine hypothesis of depression. This hypothesis proposed that a reduction in the monoamine neurotransmitters caused depression. As described in the sidebar on pages 82–83, only a small number of neurons use serotonin as a neurotransmitter, but these cells project to widespread regions of the brain. The same holds true for norepinephrine and dopamine. Although not widely used in the nervous system, these neurotransmitters are apparently involved in networks of neurons that greatly influence a person's mood. Synaptic transmission between neurons in other areas of the brain—such as neurons that process visual information, for instance—often carry specific messages, such as the presence of an object at a certain point in the person's visual field. In contrast, the monoamine neurotransmitters underlie information processing of a more general nature, some of which correlates with mood.

Because iproniazid had serious side effects, including a dangerous increase in blood pressure, researchers began looking for other antidepressant medications. Soon they found a class of drugs that became known as tricyclic antidepressants, named for the molecule's three ring-shaped structures. The first such drug was imipramine, developed in the late 1950s. Among other effects, tricyclic antidepressants inhibit the reuptake of serotonin and norepinephrine, which leaves more time for these neurotransmitters to act on the postsynaptic neuron. This mechanism is in accordance with the monoamine hypothesis.

Tricyclic antidepressants are still prescribed today, but some patients experience side effects such as dry mouth, blurry vision, constipation, and other uncomfortable conditions. Other antidepressants have since been found that induce fewer side effects. One of the most popular is fluoxetine, which is marketed under the trade name Prozac. This drug, along with Zoloft and other antidepressants, are known to inhibit reuptake proteins specifically for serotonin. As a result, these drugs are called selective serotonin reuptake inhibitors, or SSRIs. Although some concerns have appeared because of a possible risk of suicide in young patients who take Prozac, these drugs are commonly prescribed and have proved highly effective in millions of patients.

Even before the development of SSRIs in the 1980s, some researchers focused on serotonin as the most important monoamine involved

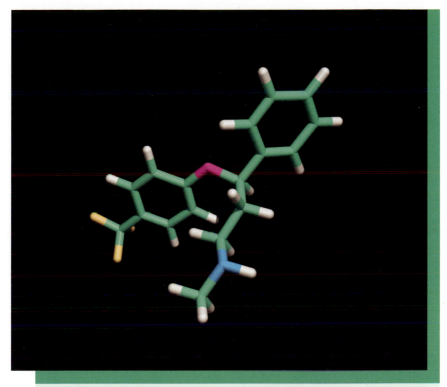

The molecular structure of Prozac—carbon atoms are shown in green, oxygen pink, nitrogen blue, fluorine orange, and hydrogen white. *(Dr Tim Evans/Photo Researchers, Inc.)*

in depression. Yet proving the assertion was difficult. For example, CSF measurements of the main metabolite of serotonin, 5-HIAA, in depressed patients showed lower concentrations in some cases, but normal in others. Physicians who tried to relieve depression by giving patients large doses of the amino acid tryptophan—from which neurons make serotonin—did not succeed. But the effectiveness of SSRIs would seem to erase any doubt—if their effectiveness could be shown to depend on their serotonin activities.

The problem is that all drugs exert a number of different effects, some of which are wanted and some of which are not (the side effects). In researching the activity of fluoxetine (Prozac), scientists noticed a discrepancy between the time the drug acts on serotonin transporters and the time at which the patient's symptoms are relieved. Many patients do not show improvement until two or three weeks after beginning Prozac treatment, yet the drug begins to affect serotonin much sooner. Kazufumi Hirano and Shizuo Yamada at the University of Shizuoka in Japan, along with their colleagues, showed that SSRIs begin to act on serotonin transporters within hours. The report, "Relationship Between Brain Serotonin Transporter Binding, Plasma Concentration and Behavioural Effect of Selective Serotonin Reuptake Inhibitors," was published in 2005 in the *British Journal of Pharmacology.*

One explanation for this delay is that the brain adjusts to the presence of SSRIs, and this adjustment is the mechanism by which the drug works. The brain may change the number of serotonin transporters, receptors, or other molecules, a process that would require time as the proteins are made and inserted in the correct positions.

Another explanation is that the monoamine hypothesis is not the whole story. In the search for alternative ideas, some researchers have noticed that new neurons generated in a brain structure known as the hippocampus may affect depression. Once thought not to occur at all in the brain of mammals, the generation of new neurons—neurogenesis—has been shown, in the 1990s by Elizabeth Gould of Princeton University in New Jersey and her colleagues, to take place in rats and primates. In 1998, Fred Gage of the Salk Institute for Biological Studies in California and his colleagues found evidence that the same is true in humans.

Testing for a possible relationship between antidepressant activity and neurogenesis, a research team led by René Hen at Columbia University in New York used mice as models. Under certain conditions, mice exhibit behavior and neural properties similar to depression in humans. Hen and his colleagues found that antidepressants, including fluoxetine, did not work if neurogenesis was blocked. This 2003 report, "Requirement of Hippocampal Neurogenesis for the Behavioral Effects of Antidepressants," published in *Science* by Luca Santarelli, Hen, and their colleagues, offers evidence of a role of neurogenesis in antidepressant activity, though no one knows if this result is also true in humans. The time required for neurogenesis could account for the delay in the effectiveness of the medications. In their report, the researchers note, "Our results suggest that strategies aimed at stimulating hippocampal neurogenesis could provide novel avenues for the treatment of anxiety and depressive disorders."

An understanding of the mechanism by which antidepressants work is important in developing other, even more effective drugs. Considering the prevalence of depression, the National Institute of Mental Health (NIMH), a U.S. government research agency, has made this research a priority. Faster-acting antidepressants would be especially welcome, avoiding the agonizing delay in recovery. As described in the sidebar on page 90, NIMH funds much of the research on the brain in the United States.

Depending on the outcome of this research, future antidepressants may or may not target serotonin or other monoamines. What is immediately clear is that while the monoamine hypothesis has prompted a lot of useful research, depression is not as simple as this hypothesis would indicate.

Another prevalent mental disorder, schizophrenia, has also had its share of various hypotheses and medications. Some of the most interesting research avenues involve the neurotransmitter dopamine.

NEUROTRANSMITTERS AND SCHIZOPHRENIA

Schizophrenia is a serious illness affecting about 1.3 percent of Americans, according to the U.S. surgeon general, and similar estimates hold for other countries. Patients suffering from schizophrenia have disorganized or irrational thoughts, often including delusions and hallucinations. The patients are also usually socially withdrawn. The Swiss psychiatrist Eugen Bleuler (1857–1939) coined the term *schizophrenia* in 1908 from Greek words meaning split mind, referring to the patient's break or split with reality.

Because of the brain's complexity, physicians and researchers have not made as much progress as they would like in efforts to understand disorders such as depression and schizophrenia. Scientists have been studying schizophrenia and related illnesses since before Bleuler's time, but as yet no one knows what causes schizophrenia. The symptoms of the disease usually make their initial appearance early in the patient's life. Since schizophrenia runs in families, there is a genetic component to the disease. It is possible that problems arise in the early stages of development, possibly due to faulty genes or perhaps in part due to exposure to environmental toxins, which fester until the disease arises in young adulthood.

National Institute of Mental Health

The idea for the National Institute of Mental Health (NIMH) was born on July 3, 1946, when then-president Harry Truman signed the National Mental Health Act. On April 15, 1949, NIMH was finally established, becoming a part of the National Institutes of Health (NIH). The mission of NIH is to support biomedical research, which it does by conducting research in its own laboratories as well as by providing funds to scientists at other institutions. There are presently 27 NIH branches, and although considerable overlap exists among these different institutes and centers, NIMH is the main supporter of brain research topics such as neurochemistry. NIMH headquarters is located at the Neuroscience Center Building in Rockville, Maryland, but many of the staff work at the NIH campus in Bethesda, Maryland.

NIMH's goal is to reduce the impact of mental illnesses and disorders by developing a better understanding of the brain and how it works. Not only does mental disease affect its victims on a personal level, the losses suffered by the economy due to reduced productivity are severe—$150 billion, according to an estimate by the Department of Health and Human Services.

NIMH fosters research with the efforts of the 500 scientists who work at the institute, and it also provides funding for scientists located at universities and research centers across the country. Scientists typically obtain NIMH funds by submitted a detailed proposal, outlining their goals and methods. Even though NIMH spent more than $1 billion on such research in 2008, funding is highly competitive—and not every submitted proposal receives funding—because there are many interesting topics that can be investigated. NIMH provided part of the funds for René Hen's research project mentioned in the text.

Another important goal is to train the next generation of scientists. Along with other NIH institutes, a portion of NIMH's budget is always earmarked to support the education of students who wish to pursue a career in science.

Prevalent theories in the early 20th century focused on psychological causes such as traumatic experiences during childhood, but much of the later research has concentrated on genetics as well as on brain abnormalities. Researchers have found the brains of schizophrenia patients to be slightly different from normal in a number of ways; for instance, the fluid-filled cavities in the brain known as ventricles are larger. But no one has found what role, if any, these differences play in the onset and time course of the disease.

A potential clue to the causes of schizophrenia came in the early 1950s, when researchers discovered the first effective drug treatment for the disease. Chlorpromazine treatments alleviated many of the symptoms, allowing some patients to recover enough function to leave hospital wards for the first time since they had been diagnosed with

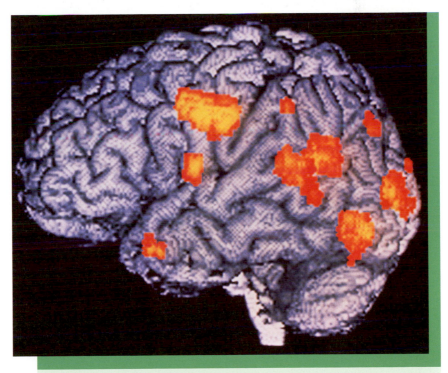

This brain scan of a schizophrenic patient, taken while he was hallucinating, shows highly active areas in the visual and auditory parts of the brain, suggesting the hallucination was the result of abnormal activity in these areas. *(D. Silbersweig/Photo Researchers, Inc.)*

schizophrenia. Investigations into the drug's activity showed that this drug binds to and blocks a certain type of dopamine receptor in the brain.

Receptors are proteins usually embedded in the cell's membrane. A neurotransmitter such as dopamine docks at its receptor, activating it and initiating some kind of response in the postsynaptic cell. The response depends on the type of receptor, and there are a number of different ones for most neurotransmitters. Researchers have found five dopamine receptors; the one affected by chlorpromazine and similar drugs is known as D_2. These drugs prevent dopamine from activating the D_2 receptor.

Although measuring brain chemicals is difficult, several techniques aid investigations of receptor function. One technique, known as autoradiography, uses radioactive versions of neurotransmitters (or other such molecules that bind receptors). Atoms of any given element have the same number of protons in the nucleus, but some of these atoms may vary in the number of neutrons and are known as isotopes of the element. Some isotopes are stable, but some are not, decaying and emitting radiation (radioactivity). For example, hydrogen-1 is a stable isotope of hydrogen—the number indicates the number of protons and neutrons in the nucleus (one proton in this case). Hydrogen-3, also known as tritium, has one proton and two neutrons, and unlike hydrogen-1, tritium is radioactive. Detectors such as Geiger counters are sensitive to the emitted radiation, as are certain chemical films.

Neurotransmitters in the body are not normally made of radioactive atoms. Researchers make radioactive versions by adding radioactive atoms to the solution in which chemical reactions produce the molecule. For example, hydrogen-3 can be incorporated into parts of a dopamine molecule instead of hydrogen-1. Brain tissue can be cut into thin slices, exposed to radioactive dopamine (or other radioactive molecules), and then examined for radioactive traces, indicating receptors to which dopamine has attached. These procedures are known as autoradiography.

In 1976, Ian Creese, David R. Burt, and Solomon H. Snyder of Johns Hopkins University in Baltimore, Maryland, reported that the most effective schizophrenia medications are the ones that have the strongest affinity for dopamine receptors. Researchers also discovered drugs that increased the amount of dopamine inadvertently caused schizophrenic symptoms in patients. These findings led to the dopamine hypothesis of schizophrenia—too much dopamine causes schizophrenia.

Like the monoamine hypothesis of depression, such a simple hypothesis was appealing but, perhaps predictably, a little too simple to be true. Further research using a technique known as positron emission tomography (PET) showed the relationship between dopamine and schizophrenia is more complex. PET detects radioactive emissions of certain isotopes; these isotopes are incorporated into a molecule and injected into a patient. The machine measures the radioactivity with detectors positioned around the body. PET lets researchers study the distribution of certain molecules in living tissue since, unlike autoradiography, the tissue is not sliced and treated chemically. The amount of radioactivity must be small, however, to avoid harming the human subjects.

Dopamine does not cross the blood-brain barrier, so researchers must inject dopaminelike molecules or other ligands, such as many schizophrenia medications that bind dopamine receptors, in order to study dopamine transmission in living subjects. PET studies confirm that schizophrenia medications block D_2. But these medications do not work in every patient, and Adam Wolkin of the New York Veterans Administration Medical Center and his colleagues discovered that dopamine receptor binding was similar in patients that respond to the medication as well as in those who fail to show improvement. The paper "Dopamine Blockade and Clinical Response: Evidence for Two Biological Subgroups of Schizophrenia," published in 1989 in the *American Journal of Psychiatry,* showed there may be a group of schizophrenia patients whose disease has a slightly different nature.

Many researchers have come to believe that schizophrenia is a complex disease, possibly with a number of different causes or courses. PET studies of schizophrenia have found possible contributions of other receptors, including the dopamine D_1 receptor as well as receptors for other neurotransmitters such as glutamate. Genetic researchers are searching for the genes involved in the expression and regulation of these receptors, any or all of which may be involved in some number of patients.

Recent improvements in chemical detection may also help researchers investigate this disease. The Duke University scientist Rima Kaddurah-Daouk and her colleagues are developing techniques to analyze hundreds of chemicals quickly and efficiently. Although many chemicals, including dopamine and other neurotransmitters, do not cross the blood-brain barrier, an accurate measure of a large number of metabolites in the blood might let researchers see the "big picture." This technology involves running the blood sample over an array of small detectors, each geared for a certain metabolite. In a paper published in 2007 in *Molecular Psychiatry,* "Metabolomic Mapping of Atypical Antipsychotic Effects in Schizophrenia," Kaddurah-Daouk and colleagues sampled the blood of 50 schizophrenia patients before and after drug treatment. Examining 300 different lipid metabolites, the researchers found differences that depended on the drug the patient received.

Technologies that examine a large number of metabolites at the same time are sometimes known as metabolomics. This term is similar to genomics, in which large numbers of genes are detected or studied simultaneously. Such studies are only in their beginning phases, and researchers do not yet understand if the differences they are finding in metabolites are critical or not. Dopamine and other brain chemicals are important in the development of schizophrenia, at least for some forms of the disease, but scientists do not yet know how.

NEUROTRANSMITTERS AND DRUG ABUSE

An important clue to dopamine's function in the brain became evident when scientists began to study the biology of drug abuse and addiction. Dopamine is one of the prominent factors.

Drugs such as cocaine (which comes from the coca plant) and morphine (which comes from the poppy plant and is the active ingredient of opium) have long been used for pain relief. But this valuable service can be spoiled by addiction—a compulsion to use the drug in excessive amounts. In the late 19th century, these drugs were available to the public without prescription—the soft drink Coca-Cola even contained small amounts of cocaine. But as abuses mounted, the United States passed the Harrison Act in 1914, regulating the use of narcotic substances. (The act was named after New York Representative Francis Harrison, who proposed the legislation.) Today, only licensed physicians may dispense drugs such as these. And, of course, sodas have milder ingredients.

Morphine activates a special synaptic transmission system that uses certain peptides, but other neurotransmitters, including dopamine, are also involved. Many narcotics, especially cocaine and amphetamine (a

Cocaine dissolves in the mucous lining of the nasal passages and quickly passes into the bloodstream, where it is carried to the brain. The user experiences the effects of the drug in a few minutes. *(Jon Schulte/iStockphoto)*

stimulant), act directly on dopamine transmission. Both cocaine and amphetamine block reuptake of dopamine, which means that more dopamine is available to activate postsynaptic receptors. Due to the links between excessive dopamine and schizophrenia, it comes as no surprise that the use of cocaine, amphetamine, and similar drugs such as "crack" (a form of cocaine) often lead to symptoms strongly resembling schizophrenia.

What is dopamine's role in drug abuse? People apparently start using substances such as cocaine because it makes them temporarily feel good—they get a "high." Part of the reason this happens is that dopamine transmission is a critical component in the neural mechanisms of reinforcement. Reinforcement encourages behavior by making the person feel good, or by removing negative feelings. Neural reward systems in the brain are important in reinforcing behaviors such as eating and mating, which are necessary to support the individual's survival as well as the survival of the species. In addition, humans respond to more complex rewards such as scholarly or athletic achievement and social interaction.

Drugs such as cocaine hijack the reward system. Cocaine makes a person "high" by stimulating the transmission normally involved in feeling good for other, more legitimate reasons. But the price paid for these temporary feelings is also high, so to speak. As users consume more of the drug, the brain adapts to the altered circumstances, reducing dopamine transmission in an effort to return to normal. This regulation may involve the amount of dopamine released, the number of receptors, or both, and plays a role in the response to psychiatric medications. Regulation also strongly influences the brain's response to repeated drug use. Users find they have to take more of the drug to get the same effects, a phenomenon known as tolerance. Another result of the adaptation occurs if users abruptly stop consuming the drug, in which case the system becomes unbalanced and users experience strongly negative feelings—these are withdrawal symptoms. To avoid withdrawal, users continue the habit, even though the drug may no longer make them feel good.

Drug abuse is a vicious, downward spiral, invariably leading to disaster. While many people manage to avoid the trap and stay away from these drugs, some people seem disposed to use and abuse these substances, and they have great difficulty escaping. Some people never escape. In order to understand why this happens, researchers are studying how dopamine neurotransmission influences and rewards behavior. Rats are often used in these studies—rats, like other animals, become addicted when exposed to drugs such as cocaine and will perform tasks to obtain the drug even as their health deteriorates.

Jeffrey W. Dalley, at the University of Cambridge in the United Kingdom, along with his colleagues, studied rat behavior in combination with chemical analyses and PET. (PET works with laboratory animals as well as with humans.) Of particular interest is an area of the brain called the nucleus accumbens, a region deep in the brain that receives a projection from neurons that use dopamine as a neurotransmitter. Drug use significantly increases this transmission.

An important question is whether the tendency to engage impulsively in risky behavior can predict who might be more susceptible to drug use. Dalley and his colleagues found that rats which show reckless or impulsive behavior—for instance, choosing the instant gratification of a minor reward for completing a task instead of a larger but delayed reward—are susceptible to addiction and also have fewer dopamine receptors in the nucleus accumbens. The report "Nucleus Accumbens D2/3 Receptors Predict Trait Impulsivity and Cocaine Reinforcement," published in *Science* in 2007, indicates that the physiological state of dopamine neurotransmission may be an important factor leading to substance abuse problems.

Dopamine and other brain chemicals profoundly influence the way people think and feel. Variability in these chemicals, particularly in the way they are regulated, affects a person's mental health, response to medication, propensity for or against danger, and other behaviors. A greater understanding of brain chemistry will not only provide more insight into what it means to be human, it will also lead to superior treatments and preventative measures to some of the most serious problems plaguing society.

CONCLUSION

Neurochemists continue to improve the methods they use to measure and analyze chemicals of the brain. They have also developed techniques to study how neurons transmit messages across synapses to other neurons, and how these messages affect the recipients. Synaptic transmission is critical in brain function, and neurotransmitter imbalances have been associated with disorders such as depression and schizophrenia as well as drug abuse. Although the cause or causes of these conditions are not necessarily as simple as too much or too little of a certain neurotransmitter, chemicals such as dopamine and serotonin are undoubtedly involved. Neurotransmitters have been excellent starting points for further, ongoing studies into these issues.

Some people wonder if researchers may one day develop procedures to regulate neurotransmission artificially. Simple procedures that influence neurotransmission already exist—including drug abuse, in which users alter their brain chemistry, albeit in an unintentional and unproductive fashion. Other substances such as naltrexone have been found that counter the "high" of drugs of abuse. Naltrexone blocks transmission of neurotransmitters known as opiates that are involved in mediating the effect of morphine and its more potent chemical derivative, heroin. In 1984, the Food and Drug Administration (FDA), which regulates food and drug safety in the United States, approved the use of naltrexone in morphine and heroin addicts. (Naltrexone should not be confused with naloxone, a drug used to counteract morphine and heroin overdoses.) Since it impedes transmission, naltrexone helps to block the rewarding effects of the drug, which discourages its use. It also helps in cases of alcohol addiction (alcoholism), for which the FDA gave its approval in 1995.

Substances such as naltrexone may help, but of course this treatment requires compliance by the patient—naltrexone must be taken in order to do any good. This prompts questions about how forcefully authorities should intervene in such cases. *A Clockwork Orange*, a 1971 film based on a 1962 novel by Anthony Burgess, involves a future society in which criminals are conditioned in ways that make them violently ill when they even consider committing a crime. Although the conditioning works, the main character in the story is unable to defend himself when he is assaulted—the very thought of violence leaves him incapacitated—and the audience is left wondering if this sort of conditioning is the best solution to the problem of criminal behavior.

Similar issues arise in the widespread use of drugs such as Prozac. The psychiatrist Peter Kramer in his 1993 book *Listening to Prozac* describes cases in which patients claim that taking Prozac brings out their "true" self. The question becomes whether Prozac is in some cases changing a person's personality instead of fixing any perceived and possibly nonexistent disorder. Such cases have led to concerns about excessive use of these drugs, as well as the development of other "medications" that do not alleviate symptoms of a disease but instead make the consumer artificially happy or contented.

It is not out of the question that further research in brain chemistry could lead to an increase in ethically or morally troubling issues. But the same complexity that daunts scientists who study the brain also works against the possibility that such troubling scenarios will come about, at least in the near future. The brain's chemicals, along with the mechanisms that regulate their use, have developed over an exceedingly long time. Human neurochemical activity is built upon a foundation perfected by evolution, and it shares many properties with animals such as the laboratory rats. If researchers have learned anything about the brain in the last few decades, it is that simple concepts—such as the monoamine hypothesis of depression and the dopamine hypothesis of schizophrenia—are rarely adequate.

Drugs that affect the brain must act in complicated ways as well. These complications mean, at least in the author's opinion, that there will always be side effects that limit a drug's usefulness and effectiveness—including any present or future "lifestyle" drugs that supposedly alter personalities. Brain chemistry is a frontier of science in which research has much promise to understand, and possibly fight, mental disorders and abusive behavior, and the risk of this knowledge being abused in some way is not nearly as great as the potential reward.

CHRONOLOGY

1719 C.E.   German scientist Johann Hensing (1683–1726) makes one of the first attempts to chemically analyze the brain. His analysis is simple, though he does identify elements such as phosphorus.

1791   French scientist Michel-Augustin Thouret (1748–1810) makes another early attempt at a chemical analysis of the brain, though his crude analytical methods yield only soapy materials fats.

1811   French chemist Louis-Nicolas Vauquelin (1763–1829) improves the techniques to chemically analyze the brain and finds unusual fatty materials containing phosphorous.

1833   French researcher and psychiatrist Jean-Pierre Courbe claims that phosphorus in the brain correlates with excitability—too much phosphorus is associated with madness and too little with mental deficiency. His analysis, though mistaken, heralds similar conclusions concerning other brain chemicals.

1873   Italian physician Camillo Golgi (1843–1926) develops a staining technique that vastly improves researchers' ability to see and study neurons.

1897   British physiologist Sir Charles Sherrington (1857–1952) coins the term *synapse.*

1906   Camillo Golgi and Spanish anatomist Santiago Ramón y Cajal (1852–1934) share the Nobel Prize in physiology or medicine for their research on the brain. Although Golgi continues to argue that neurons may be physically connected, Cajal correctly states that most neurons are separate cells.

British researchers Thomas Elliot (1877–1961) and John Langley (1852–1925) develop the idea of a receptor—a molecule to which chemicals bind and exert effects on a cell.

1914   The U.S. government passes the Harrison Act, which regulates narcotic substances.

1921   German-born physician Otto Loewi (1873–1961) performs an experiment with frogs in which he stimulates a frog's vagus nerve, a branch of which influences heart rate. When Loewi transfers the fluid surrounding this vagus nerve onto the isolated heart of another frog, its rate is affected in the same way. This experiment suggests that the vagus nerve releases some substance that acts on its target organs.

1934   German physiologist Wilhelm Feldberg (1900–93) and British researcher Sir John Gaddum (1900–65) show that chemical transmission takes place, demonstrating that certain nerves release the chemical acetylcholine.

1948   Maurice M. Rapport, Arda Alden Green, and Irvine H. Page of the Cleveland Clinic Foundation in Ohio isolate and name the chemical serotonin, which they discover in the blood.

1950s   The monoamine hypothesis of depression develops from the finding that drugs reducing monoamine neurotransmission cause depression. Monoamines include serotonin, norepinephrine, and dopamine.

1957   Swedish chemist Arvid Carlsson discovers that dopamine is a neurotransmitter.

1966   Dutch researcher Jacques van Rossum proposes the dopamine hypothesis of schizophrenia, which attributes the symptoms of the disease to an excess of dopamine.

1980s   Selective serotonin reuptake inhibitors (SSRIs) such as fluoxetine (Prozac) begin to be used as antidepressants. These medications are generally effective and have fewer side effects than earlier drugs.

               Researchers begin to examine dopamine's role in the brain's reward system and in the reinforcing aspects of drugs of abuse.

1990s   Although acknowledging the relevance of monoamine neurotransmitters in depression, researchers begin to examine other mechanisms such as neurogenesis that may also play an important role in the development of the disease.

               Improved methods of studying brain chemicals, including methods such as PET that permit the study of living tissue, confirm the importance of dopamine in schizophrenia, though it is probably not the only factor. Scientists study other factors, many of which are related to synaptic transmission, such as genes involved in synaptic transmission.

2003   Luca Santarelli, René Hen, and their colleagues use a mouse model of depression and find evidence of a role of neurogenesis in antidepressant activity.

2007   Jeffrey W. Dalley and his colleagues use animal models and imaging to show that the physiological state of dopamine neurotransmission may be an important factor in substance abuse problems.

FURTHER RESOURCES
Print and Internet

Braun, Stephen R. *The Science of Happiness: Unlocking the Mysteries of Mood.* New York: Wiley, 2001. Mood disorders and the drugs prescribed for them are often controversial—when does a normal bout of sadness bloom into a depression that needs medication? Braun is a journalist who investigates this issue from a variety of perspectives.

Chudler, Eric H. "Neurotransmitters and Neuroactive Peptides." Available online. URL: http://faculty.washington.edu/chudler/chnt1.html. Accessed May 28, 2009. Eric Chudler is a professor at the University of Washington in Seattle, Washington. This Web resource is part of his excellent Web site, Neuroscience For Kids, and contains a nicely illustrated tutorial on neurotransmission.

Dalley, Jeffrey W., Tim D. Fryer, Laurent Brichard, Emma S. J. Robinson, David E. H. Theobald, Kristjan Lääne, et al. "Nucleus Accumbens D2/3 Receptors Predict Trait Impulsivity and Cocaine Reinforcement." *Science* 315 (March 2, 2007): 1,267–1,270. The researchers report on evidence that indicates the state of dopamine neurotransmission may be an important factor leading to substance abuse problems.

Hirano, Kazufumi, Ryohei Kimura, Yumi Sugimoto, Jun Yamada, Shinya Uchida, Yasuhiro Kato, Hisakuni Hashimoto, and Shizuo Yamada. "Relationship Between Brain Serotonin Transporter Binding, Plasma Concentration and Behavioural Effect of Selective Serotonin Reuptake Inhibitors." *British Journal of Pharmacology* 144 (2005): 695–702. The researchers show that SSRIs begin to act on serotonin transporters within hours.

Howard, Pierce J. *The Owner's Manual for the Brain: Everyday Applications from Mind-Brain Research,* 3rd ed. Austin, Tex.: Bard Press, 2006. With scientific research as its starting point, this book discusses how the present state of knowledge in brain science can have a practical impact on human lives. A wide variety of topics are discussed, including brain chemistry, the effects of drugs, brain disorders, learning and memory, and much else.

Kaddurah-Daouk, R. J. McEvoy, R. A. Baillie, D. Lee, J. K. Yao, P. M. Doraiswamy, and K. R. R. Krishnan. "Metabolomic Mapping of

Atypical Antipsychotic Effects in Schizophrenia." *Molecular Psychiatry* 12 (2007): 934–945. Kaddurah-Daouk and colleagues sampled the blood of 50 schizophrenia patients before and after drug treatment and found differences in metabolites that depended on the drug the patient received.

Kramer, Peter D. *Listening to Prozac.* New York: Penguin, 1993. Kramer, a psychiatrist, describes the profound effects of the popular antidepressant Prozac in his patients.

LeDoux, Joseph. *Synaptic Self: How Our Brains Become Who We Are.* New York: Penguin, 2003. LeDoux, a researcher at New York University, examines some of the most fundamental issues of brain science, such as consciousness. LeDoux is a fine writer, and included in the discussion is his interesting perspective on the biology and chemistry of mental illness.

Ramón y Cajal, Santiago. Nobel Lecture, 1906. Available online. URL: http://nobelprize.org/nobel_prizes/medicine/laureates/1906/ cajal-lecture.pdf. Accessed May 28, 2009. Santiago Ramón y Cajal's Nobel lecture describes his research techniques and results.

Santarelli, Luca, Michael Saxe, Cornelius Gross, Alexandre Surget, Fortunato Battaglia, Stephanie Dulawa, et al. "Requirement of Hippocampal Neurogenesis for the Behavioral Effects of Antidepressants." *Science* 301 (August 8, 2003): 805–809. The researchers found that neurogenesis is important in antidepressant activity in mice.

University of Texas Addiction Science Research and Education Center. "Understanding Addiction: Basic Science Information." Available online. URL: http://www.utexas.edu/research/asrec/addiction.html. Accessed May 28, 2009. This Web page provides an abundance of information on neurons, neurotransmitters—especially dopamine— and drugs that influence the transmission process.

Wolkin, Adam, Faouzia Barouche, Alfred P. Wolf, John Rotrosen, Joanna S. Fowler, Chyng-Yann Shiue, et al. "Dopamine Blockade and Clinical Response: Evidence for Two Biological Subgroups of Schizophrenia." *American Journal of Psychiatry* 146 (1989): 905–908. The researchers discovered evidence for two different types of schizophrenia.

Web Sites

National Institute of Mental Health: Depression. Available online. URL: http://www.nimh.nih.gov/health/topics/depression/. Accessed May 28, 2009. The National Institute of Mental Health, a branch of the National Institutes of Health, supports basic and applied research in a variety of fields in brain science and mental health. This Web site supplies information on the nature of depression, its signs and symptoms, and treatment options.

National Institute of Mental Health: Schizophrenia. Available online. URL: http://www.nimh.nih.gov/health/topics/schizophrenia/. Accessed May 28, 2009. The National Institute of Mental Health maintains this informative Web site on schizophrenia. Topics include the nature of the disease, signs and symptoms, and treatment options.

Schizophrenia.com. Available online. URL: http://www.schizophrenia. com/. Accessed May 28, 2009. Schizophrenia.com is a nonprofit Web-based community devoted to providing information and support to schizophrenia patients and their families. Their Web site contains a huge array of information, including news and discussion groups.

4

SMART MATERIALS— MATERIALS THAT ADAPT TO CHANGING CONDITIONS

Imagine a peregrine falcon soaring high in the sky. To maintain its altitude, the bird extends its wings to their full 40-inch (100-cm) wingspan, maximizing the lift. Then its keen eyes spot a dove flying below. The falcon draws in its wings, sweeping them back into a shape resembling a sickle. A steep dive ensues, known as the "hunting stoop," in which the peregrine falcon can reach an astounding speed of 200 miles per hour (320 km/hr). If the hunt is successful, the falcon smashes into a wing of its target. The dove goes down, and the falcon retrieves its prey.

Unlike the wings of most airplanes, bird wings are not fixed. Peregrine falcons change the shape of their wings to suit two different needs—soaring and diving. The outstretched wings provide lift, which is excellent for staying aloft but a bad choice for diving. By sweeping the wings back, peregrine falcons plummet to the ground in a fast and controlled dive. Then, when the birds want to soar again, the wings can be extended.

The wings of peregrine falcons offer an important lesson to chemists and other scientists who develop new materials. Most of the materials described in the previous chapters involve discovering or designing new molecular structures to meet a prescribed need. These materials fulfill the

Peregrine falcon *(Wildlife/Peter Arnold, Inc.)*

need by excelling at one particular type of job, such as binding to a specific chemical or conducting a small amount of electrical current. If the job changes, or the conditions under which the job is to be performed changes, the material is no longer optimal and may not function at all. To extend its performance range, a material must change and adapt—to be "smart" rather than remaining static or fixed.

Smart materials are extremely efficient, performing a variety of jobs that would otherwise require researchers and engineers to spend a lot of time and money to develop and design a number of different substances. Materials capable of multitasking are one of the most interesting and promising frontiers of chemistry and materials science. This chapter describes research on a variety of materials and systems that exhibit some degree of responsiveness to their environment.

INTRODUCTION

Which material is best for any given situation depends on the job requirements. In airplanes and other flying vehicles, for example, the frame and wings must be made of a strong material capable of safely supporting the passengers and withstanding the powerful forces of high-speed winds. Yet the material should be as lightweight as possible since heavy vehicles are difficult to get off the ground and consume much more fuel. The first airplane to make a sustained flight, the Wright Flyer of 1903, built by Orville (1871–1948) and Wilbur Wright (1867–1912), was made of wood (spruce and ash) covered by a cotton fabric known as muslin. The craft was relatively strong but also light and flexible.

To control their flying machine, the Wright brothers knew they would have to adjust the lift of the wings. Lift is created by the flow of air around the wing; the shape of the wing and the angle it makes with respect to the flowing air are critical. The Wright brothers used wires to warp the wings of their 1903 Wright Flyer, adjusting the wing shape in order to steer and control the flight. This and other control mechanisms of the aircraft were primitive, and the plane was not easy to guide. As described at eyewitnesstohistory.com, Orville Wright explained in his diary: "I found the control of the front rudder quite difficult on account of its being balanced too near the center and thus had a tendency to turn itself when started so that the rudder was turned too far on one side and then too far on the other. As a result the machine would rise suddenly to about 10 ft. and then as suddenly, on turning the rudder, dart for the ground." The first flight lasted only about 12 seconds.

The Wright Flyer's flexibility permitted the wing-warping design, yet was strong enough to endure the forces encountered on the first flights, on December 17, 1903, since the speed was only around 5–10 miles per hour (8–16 km/hr)—hardly enough to impress a peregrine falcon. But as engines became more powerful, airplane speeds increased. Winds and air resistance encountered at high speed would tear apart flexible wings, so aviation engineers had to strengthen them. Wings became rigid instead of flexible. In order for the pilot to control the airplane, manufacturers added stiff but adjustable surfaces such as flaps and ailerons. Most of the fighter planes of World War I (1914–18), flown by pilots such as German ace Manfred von Richthofen (the "Red Baron") (1892–1918) and American ace Eddie Rickenbacker (1890–1973), were made of wood.

Although wood is a good material, metal and mixtures known as alloys are stronger. Early metal airplanes were made of iron or steel, but in the early 1920s, German manufacturer Hugo Junkers (1859–1935) made an all-metal plane with a skin of an aluminum alloy called duralumin. (Although expensive, aluminum was used earlier for certain components—the Wright brothers used an engine made with aluminum in their 1903 airplane.) Aluminum alloys are ideal for flying since they fulfill two of the most important requirements—they are strong yet lightweight.

Atoms of a metal or alloy can slide around a little bit, even though the bonds holding them together are strong. This flexibility is useful because the material can be flattened or rolled into shape. Yet the strength of the bonds results in a hard material, capable of withstanding a lot of force. Because an alloy is a combination of different elements, it often tends to be harder than a pure metal—the differently sized atoms making up an alloy tend not to slide or move against one another as easily as the identical atoms of a pure metal. Aluminum, with an atomic number of 13, is not massive, though in its pure form it is quite soft. But when combined with copper, magnesium, manganese, or other elements, aluminum makes strong alloys. Although aluminum alloys are not generally as strong as steel, they are not all that much weaker and have about 1/3 the weight.

Most of the airplanes of today, and space ships such as the space shuttle, are made of aluminum. But engineers are increasingly using composites instead of aluminum. Composites, as described in chapter 1, are composed of thin fibers reinforcing a matrix of plastic or metal. Although these materials are generally expensive and require a little more care than aluminum alloys, they offer an excellent combination of strength and lightness of weight. About 30 percent of military aircraft are now made from composites.

But the most efficient material is one that adapts to changing conditions. Consider the iris and pupil of the eye, for example. The iris—the colored part of the eye—has a hole, called the pupil, which lets light into the eye. This structure is sufficient for vision, but if it were not able to change, it would be far from ideal. At night, the eye needs a greater opening to let in more of the scarce light; at noon on a sunny day, the opposite problem occurs—letting in too much light would overload the sensitive retina, the part of the eye that converts light into electrochemical signals. For optimal functioning, a pupil should get larger at night and smaller during the day. Thanks to muscles embedded in the iris, this is exactly what happens.

Just as varying levels of light place different demands on the eye, varying circumstances pose different challenges to the variety of available engineering materials. Soaring and diving during flight are best served by different wing shapes, a problem that the peregrine falcon's flexible wings solve. Fixed-wing aircraft are not as flexible. And take-offs and landings, which are critical moments in a flight where many accidents occur, offer entirely different problems—take-offs require much lift, but landings do not.

Any single solution to a problem requiring opposite properties is bound to be a compromise—adequate for both conditions but optimal for neither. Most aircraft wings have attached surfaces, which when moved back and forth lengthen or shorten the wing area or adjust some other property. These movements allow pilots to control the plane, but the moving parts provide only limited flexibility, and they add a lot of weight and complexity to the craft, resulting in a serious loss of ease and efficiency. A few bold aviation engineers have designed and built airplanes with movable wings, such as a military jet called the F-14 Tomcat, which can sweep its wings back and forth. Such designs offer a little more flexibility, but they are poor imitators of the peregrine falcon.

The ideal solution is a material that can change its shape or properties as needed. A single, adaptable material or structure requires no compromises.

A MATERIAL'S I.Q.— RESPONSIVENESS TO CHANGE

Most materials undergo changes when certain features of their environment change. One of the most common changes is called thermal expansion—most substances increase in volume as the temperature rises. (Heat causes a substance's atoms and molecules to increase their motion, which results in an increase in the substance's size.)

Thermal expansion is a reaction to the environment, but it is not very useful for engineering purposes. The expansion is typically small—for example, a 1,000-foot (305-m) steel beam will get only about 3.5 inches (8.75 cm) longer when the temperature increases from 50°F (10°C) to 100°F (37.8°C). Even though it is not a drastic change, engineers must take it into account, otherwise structures such as bridges and railway tracks might fail when the temperature changes too much.

Changes like thermal expansion are generally more of a nuisance than anything else. Although thermal expansion is convenient in making certain thermometers, in most cases it is not a useful adaptation.

Exploiting a material's adaptability usually requires a larger change and a more effective way of controlling it. A material's I.Q.—intelligent quotient, as whimsically used by materials scientists to describe how "smart" a smart material is—depends on the material's responsiveness. Smart materials should have a large response magnitude, achieved with agility and quickness.

For example, certain automated processes—running with little or no human intervention—require temperature regulation so that the parts do not unduly expand or otherwise change functionality. What kind of material could provide adequate temperature regulation for these automated processes? Some materials such as platinum experience a rise in electrical resistance when heated; resistance refers to an opposition to the flow of current, and an object's resistance is a property that can be measured with a great deal of precision. Platinum is a well-behaved material in this regard because its resistance varies smoothly with temperature, with no jumps or dips that would throw off the measurements. But the response, although smooth, is extremely small. The response must be used to affect some mechanism in the automated process so that any rise in temperature would be opposed or compensated, and the small response of platinum would probably not be sufficiently sensitive to launch an effective counteraction. But certain materials such as barium titanate, with a small number of other elements added, responds to even small temperature changes by increasing its resistance thousands of times. This material can be part of an electrical circuit that changes quickly and automatically when exposed to higher temperatures, providing adequate regulation.

Other materials sense and respond to different features of the environment. For example, compounds known as photochromic materials change color in response to light. (The term *photochromic* derives from Greek words meaning light and color.) In the 1960s, Corning Incorporated developed photochromic glass, which is now being used to make photochromic lenses for eyeglasses. Photochromic lenses darken when exposed to ultraviolet radiation—electromagnetic radiation of a slightly higher frequency than the violet end of the visible light spectrum—but remain transparent otherwise. The sun emits a lot of ultraviolet radiation, which in high doses can be dangerous to the skin, resulting in sunburn and other damage, as well as to the eyes. The following sidebar discusses how photochromic materials work.

Photochromic Material—Changing Color in Response to Light

Compounds that respond to light have long been known. Some of these compounds are organic—which contain carbon and are associated with living organisms—and others are inorganic. One of the most common uses of the inorganic compounds that respond to light is in photography. First developed in the middle of the 19th century, photographic film or plates generally rely on compounds known as silver halides—combinations of silver with one of the halogens (fluorine, chlorine, bromine, iodine, and astatine). Silver chloride, a popular choice, turns dark when struck by light.

The response to light that occurs in photographic films is not reversible—once an exposed film turns dark, it will not change back. Irreversibility is not appropriate for eyewear (eyeglasses that could be used only once would hardly be popular), but researchers eventually discovered a way of embedding silver halides in glass or plastic that would make them respond reversibly to electromagnetic radiation. Ultraviolet radiation modifies the atoms of silver to accept electrons from the surrounding glass, resulting in a darkening of the material. In the absence of radiation, the electrons escape, returning the material to its transparent state.

The speed and extent of the reaction depend on the number of silver atoms and the composition of the lens. Other photochromic compounds are sometimes used, though the process is similar and is also reversible.

Photochromic lenses are often used in prescription eyeglasses—eyeglasses that are worn to correct vision. Having photochromic lenses protects the eyes and, similar to sunglasses, shields the user from bright sunshine outdoors. Some prescription eyewear can be tinted like sunglasses, but if the tint does not change—as in ordinary sunglasses—then the user has difficulty indoors. Photochromic lenses avoids the need for two pair of prescription eyeglasses, one for inside and one for outside, because a photochromic lens is dark only outside.

But photochromic lenses are not perfect. Since car windshields block most ultraviolet radiation, a lot of photochromic eyeglasses will remain clear when the wearer is inside a car, even though sunglass-like darkening would be preferred while driving during the day. Even outside, the angle of the sun's ray can sometimes affect the extent of darkening, adding an undesirable variability.

To correct these problems, researcher Chunye Xu and her colleagues at the University of Washington in Seattle are developing adjustable lenses. These lenses are not photochromic—they do not respond to lighting conditions—but instead are electrochromic, which means the lenses change color in response to varying electrical conditions. An early version of these glasses uses an organic compound that changes when a small voltage is applied. Because of the great variety of organic compounds, the researchers have been able to find compounds that not only turn dark but also different colors, such as blue, red, and yellow. A battery supplies the voltage, and the wearer can adjust the shade by turning a dial.

ADAPTIVE AND INTELLIGENT SYSTEMS

Using smart materials in objects such as eyeglasses means that the function can change depending on environmental conditions or the needs of the user. Smart materials can also be critical components of systems—assemblies of interacting parts—that adapt to varying conditions. Such systems are known as adaptive or intelligent systems.

Adaptive systems contain three types of components. There must be a sensor, a component that detects or measures the varying condition. If the conditions warrant a change in the system's function, a component called an actuator will perform the adjusting. An actuator is often a motor, such as the motor in an electric drill, which causes or creates motion. The component called the controller determines when and how the system's function should change. This component is usually a small computer, which receives input from the sensor and switches on the actuator, if needed.

One important example of an adaptive system occurs in optics. Adaptive optics was developed in the 1990s and refers to the process of rapidly and automatically correcting distortions that can occur in optical systems. For instance, telescopes on the surface of Earth (unlike the orbiting Hubble Space Telescope) must view astronomical objects through the atmosphere. But air affects the path of light, as is commonly seen in mirages such as the "water" that appears in the distance as a driver travels across a long stretch of road on a sunny day. (The "water" is actually the blue sky because the warm air rising from the pavement bends light rays coming from the sky.) Distortion can ruin a telescopic image and is a source of much frustration for astronomers, which is one of the reasons most ground-based telescopes are located high on mountains—the great height means that light does not have to travel through as much atmosphere as it would if the telescope were at sea level.

But even at the thin atmosphere at high altitudes, atmospheric distortion limits the ability of telescopes to discern faint objects. Adaptive optics helps by sensing this distortion and instantly adjusting the optical properties of the instrument to counteract the harmful effects. The process must operate continually since air moves around, and the extent and nature of atmospheric distortion changes rapidly.

The W. M. Keck Observatory, located on the summit of 13,796-foot (4,206-m) Mauna Kea in Hawaii, has employed adaptive optics since 1999. Sensing the distortion requires examining the light of a bright star in the patch of sky under interest or by shining a laser beam off a thin layer of atoms high in the atmosphere (thereby creating a "virtual" star that can be examined even in regions of the sky with no bright stars). After sensitive optical instruments determine the amount of distortion, a computer makes corrections by adjusting the path that light takes through the telescope. These adjustments occur when actuators bend a six-inch (15-cm) deformable ("rubber") mirror, which is one of the mirrors that route the light through the telescope and into the imaging device. (The main mirror of each of the two Keck telescopes is much bigger—394 inches [10 m].) The small deformable mirror is made of thin glass, and 349 actuators alter its shape slightly, hundreds of times every second. Only small adjustments are necessary in optics—the Hubble Space Telescope required a repair in 1993 due to a flaw in the shape the size of 1/50th the diameter of a human hair—and the actuators bend the deformable mirror by similar distances. But the small adjustments are enough to allow the Keck telescopes to generate images that are 10 times sharper and perform nearly as well as the Hubble Space Telescope under certain conditions, despite the effects of the atmosphere.

RESPONDING TO ELECTRICITY AND MAGNETISM

Light is electromagnetic radiation, and electricity and magnetism, along with optics, are extremely important in science, engineering, and industry. Smart materials and adaptive systems that respond to some aspect of electricity and magnetism are in high demand. A circuit based on electrical resistance was mentioned earlier in the chapter.

Materials that respond to electricity or magnetism were some of the earliest smart materials to be discovered. The British researcher James Joule (1818–89) found in 1842 that iron changes length in response to a magnetic field, a process called *magnetostriction*. Since electric and magnetic fields are easy to produce and control with precision, these smart materials can be extremely useful.

Piezoelectric materials generate a voltage when pressed or squeezed. (The name for this group of materials comes from a Greek word, *piezein,* which means to press.) Voltage is an electric potential—an ability to cause a current to flow—and in piezoelectric materials, the voltage arises when the pressure causes electric charges to line up or separate. The opposite effect occurs as well; when a voltage is applied to a piezoelectric material, it compresses. These voltages and movements are small (the compression is generally a fraction of 1 percent of the total length), but reliable. Common piezoelectric materials include quartz (silicon dioxide) and Rochelle salt (sodium potassium tartrate—the name Rochelle salt comes from the French town La Rochelle, where it was first separated).

But responses of small magnitude are not necessarily useful. Is there a way of increasing the magnitude? An analogy with alloys may be helpful—combining atoms of different sizes, as in mixing elements to make an alloy, often results in a harder material because the atoms do not slide past one another so easily.

Chemical engineers can sometimes increase the size or alter the nature of the response of a smart material by a process known as doping.

Doping is an unattractive term for the introduction of a trace amount of other elements or substances into a material. For example, one of the most commonly used piezoelectric materials is lead zirconate titanate, abbreviated PZT (the *P* stands for Pb, the chemical symbol of lead, which derives from the Latin word for lead, *plumbum*). PZT contains lead, oxygen, titanium, and zirconium. Adding a tiny amount of an element such as niobium introduces some slight instability in the material's structure—the niobium atoms infiltrate the structure and alter its geometry in a few places. These instabilities are like "tipping points" or places where the structure is not quite "balanced," and an applied electric or magnetic field can initiate more pronounced changes than it could for a stable, steady structure.

Piezoelectric materials make good sensors, and movements induced by applying voltages to piezoelectric materials form the basis for a lot of actuators. These movements can be quickly and accurately controlled with electrical devices such as computers. Adaptive systems require such components. Actuators in deformable mirrors used in many adaptive optics systems are piezoelectric. (The actuators in the deformable mirror at Keck Observatory, however, operate by a similar though slightly different principle known as electrostriction: Nonconducting materials—materials that can carry little or no electric current—commonly change shape under the application of an electric field. But unlike the piezoelectric effect, electrostriction is not proportional to the applied voltage, nor does it reverse the response when the sign—plus or minus—of the voltage is reversed. Electrostrictive actuators generally have a faster response time, though the responses are more complicated.)

Piezoelectric materials are also used in alarm systems and medical imaging such as ultrasound. In some ultrasound machines such as those used to image the internal organs of a patient or the fetus of a pregnant woman, the high-frequency waves come from applying alternating high-frequency electric fields to a piezoelectric crystal. This causes the crystal to vibrate, pushing against air and producing high-frequency sound waves (which are beyond the range of human hearing, and are therefore called ultrasound).

Magnetostriction, mentioned above, was an early discovery. As illustrated in Part A of the figure, application of a magnetic field, such as that created by a powerful magnet, induces the tiny magnetic particles in a magnetostrictive material to align, changing the material's

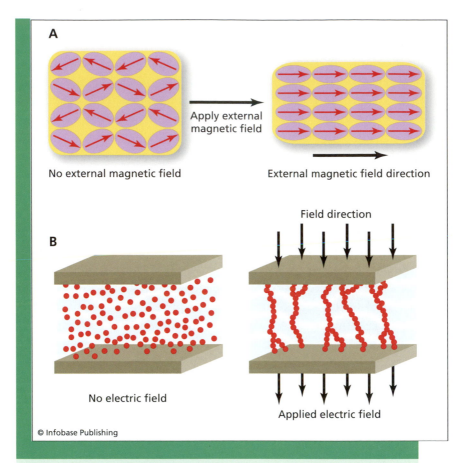

A

Apply external
magnetic field

No external magnetic field

External magnetic field direction

Field direction

B

No electric field

Applied electric field

© Infobase Publishing

(A) As a result of applying a magnetic field, the poles of tiny magnetic particles in a magnetostrictive material line up, which alters the shape of the material. (B) The application of an electric field aligns particles in an electrorheological fluid.

shape. This alignment is similar to the manner in which a magnet will align iron filings. The opposite effect—a pressure-induced magnetic field—also occurs in these materials. Magnetostriction occurs in materials that possess tiny magnetic particles, or domains, such as iron and nickel. Alloys of these elements are common magnetostrictive materials. The size of the effect is small, only a few parts per million, except in rare cases.

Applications of magnetostriction include sensors and actuators. Magnetostrictive materials can perform many of the same functions as piezoelectric materials, though specific properties vary. Production of ultrasound, for instance, can be done with magnetostriction, but this effect tends to be limited to lower frequencies than piezoelectric materials.

Other electric and magnetic effects involve fluids. Rheology refers to the deformation or flow of materials, and electrorheological fluids are substances that change the way they flow when an electric field is applied. Magnetorheological fluids do the same under a magnetic field. An important property of fluids is their viscosity, which is a measure of how easily it flows. A highly viscous liquid such as tar or pitch is thick and flows slowly, whereas a liquid with low viscosity such as water is thin and flows freely. In electrorheological and magnetorheological fluids, application of an electric field or a magnetic field, respectively, significantly alters the viscosity.

Part B of the figure on page 116 illustrates the process. These fluids typically contain a suspension of small particles, with sizes that are as big as half the diameter of a human hair on down to the nanoscale. The applied field aligns them or causes them to interact in such a way that they form chains or columns. As a result, the liquid has a higher viscosity and flows less freely, among other changes. The effect can be dramatic, changing a thin, runny liquid into a substance with a thick, gel-like consistency almost instantaneously. In electrorheological fluids, the suspended particles are commonly metal oxides, silica, or organic compounds, whereas magnetorheological contain particles such as iron shavings that respond to magnetic fields. The fluid is often some type of oil.

Smart fluids, as they are sometimes called, are subject to settling of the particles but have become increasingly used in engineering and industry. Applications involving a damping or controlling motion are particularly important. These conditions occur, for example, when a car speeding down the interstate highway hits a bump. A suspension system is needed to absorb the impact; if there was no suspension system, the driver's rear end and spine would do the absorbing, making for an extremely uncomfortable ride.

Although springs in the car's suspension system take the hit, they tend to oscillate if their motion is not damped. Damping is the job of shock absorbers. But a shock absorber must operate under varying road conditions, from new and glassy asphalt to an old road full of potholes. Conventional shock absorbers work hydraulically—using fluid pressure and valves—but in 2002, car and truck manufacturer General Motors began introducing shock absorbers based on magnetorheological fluids in a few of their models. These shock absorbers are expensive, but the fast-acting smart fluids adjust more efficiently and effectively to varying road conditions and vehicle speeds. According to a 2003 General Motors product description, "[T]his design offers superior handling, control and ride quality on the roughest road surfaces because it automatically minimizes damping forces as needed for improved road isolation and ride smoothness. It can respond to inputs in one millisecond, or 10 times faster than systems on the market today."

SHAPE MEMORY ALLOYS

Imagine a material smart enough to "remember" its original shape. If it gets deformed, it can regain its former structure with the application of a little bit of heat. The figure opposite illustrates this process.

A type of material known as shape memory alloy (SMA) can perform this trick. SMAs are more complicated than electrorheological fluids and the other smart materials previously described in this chapter. An SMA does not only react or respond to environmental conditions, it also has a memory that enables it to return to a specific structure, or sometimes switch between two different structures. After the material has been set, it can recover from a deformation that would be permanent in other materials. When the temperature is raised by an amount that depends on the specific material, it snaps back into shape automatically. The memory is based on phase transitions, as described in the sidebar on page 120.

The most common SMAs are nickel-titanium alloys and copper alloys of various kinds. Nitinol, a specific alloy of nickel (Ni) and titanium (Ti), is probably the most widely used. (The word *nitinol* comes from the chemical symbols of its two metal components, along with an abbreviation for the Naval Ordnance Laboratory, where this alloy was discovered and studied in the early 1960s.) Although nickel and titanium alloys tend to be more expensive than copper materials, nitinol has a better memory and can recover from more severe deformations.

In addition to memory, SMAs have another, related property that enables them to snap back from loading strains. Elasticity is a property of a material in which it returns to its original shape after being bent or dented by a pressure, as from carrying a load, for instance. A rubber ball is elastic—a finger poking the surface will create a dent, but the ball regains its round shape when the finger is removed. Many materials are elastic, at least to a certain extent, though all such materials have a limit beyond which they cannot recover, in which case the material breaks or is permanently deformed.

Stress created by a load will cause an SMA to deform, but if the temperature is high, it will also cause the material to transition into the soft, deformable phase. This process is similar to the transition that occurs from high to low temperatures, as discussed in the sidebar on page 120, except in this case the material started out in the high-temperature phase and it is the stress that causes the transition—the temperature does not need to change. When the load is removed, the SMA automatically regains its original shape, unless the resulting strain was too much. The advantage in using an SMA over other materials is that SMAs can recover from greater strains. Nitinol can bend about 15–20 times more than steel before breaking.

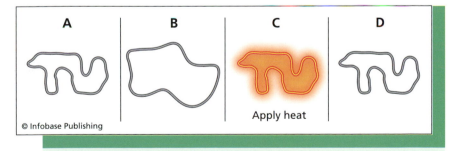

(A) Wire made of a special kind of smart material has been fashioned into a certain shape. (B) The wire gets bent out of shape. (C) Heat is applied, elevating the material's temperature. The wire automatically returns its original shape. (D) The shape is maintained after the material cools.

Phase Transitions and Shape Memory Alloys

In chemistry, a phase transition describes a change in the phase, or state, of a substance. Common phases are gaseous, liquid, and various solid phases such as crystalline, in which the atoms and molecules are configured in a repeating geometric pattern (a cube, for example). Phase transitions occur when the substance goes from one phase to another, as when ice melts into water, or water evaporates into vapor. Temperature is an important factor that can cause these changes because it stirs atoms and molecules, breaking bonds that hold them together. Ice melts, for example, when heated. In the opposite case, a drop in temperature allows the bonds to form, so that water becomes ice.

Some substances undergo transitions between phases even though the outward appearance does not change. Certain solids have multiple phases consisting of different molecular arrangements; a transition between these two phases results in considerable internal rearrangement, though the material remains a solid.

An SMA has a number of different phases as a solid, one of which occurs at high temperatures. Suppose the material is heated and shaped. When the SMA cools, it adopts another molecular arrangement, remaining a solid and retaining its shape but becoming softer and more deformable. The amount of cooling depends on the material, but it is usually not many degrees—perhaps a change of 20°F (11.1°C). If deformed at this cooler temperature, the SMA maintains this new shape. But heat causes the atoms to return to the high-temperature phase. As a consequence of this phase transition, the material snaps back to its original shape when heat is applied.

There is a great need for strong materials such as alloys that can snap back into shape. Medical applications include prostheses—artificial limbs—and implanted devices such as heart valves. Most biological substances are "smart," and the ability to replace lost or injured tissues and organs with smart materials would be a tremendous medical advance.

Some researchers are already making progress. Consider heart valves, for example. A heart valve consists of flaps of hard tissue controlled by muscles. The heart has four chambers, and four valves open and close to keep the heart contraction from squirting blood through the wrong vessel. (The opening and closing of these valves create the sounds of a heartbeat.) In some people, the valves do not work correctly due to a defect or injury. Some of these patients need surgery in which one or more valves are replaced, usually with valves from a human donor or perhaps taken from an animal (usually a pig), or a mechanical valve.

Valve replacements do not always last a long time, and they are complicated in children because the valves do not grow with the child. Children may require several surgeries as they age, which are risky and require lengthy recuperation. But in 2006, Gregory Carman and Lenka Stepan at the University of California, Los Angeles (UCLA) and Daniel Levi at UCLA's Mattel Children's Hospital, designed and built a heart valve made with nitinol. The valve, consisting of a thin film of nitinol, can be folded up and placed into a tube called a catheter. Physicians insert the catheter into a blood vessel and carefully slide it into the heart. When the nitinol valve is in place, the physician releases a catch and the valve, made of shape-memory alloy, remembers its original shape and unfolds, yet retains the strength necessary to function at the demanding pace of about 90,000 beats a day.

Catheter procedures avoid the need for surgeons to open the patient's chest. As a result, the patient experiences a lot less pain and needs far less time for recovery. These procedures have already become common in adults, but the bulky valves and catheters are not suitable for children. With nitinol's memory and flexibility, the valve can be fitted into a small space yet return to its needed size when placed in position. Nitinol is biocompatible—it does not harm tissues—so it causes no damage when implanted in the body.

Components of some cell phones contain shape memory polymers.
(Jerry Mason/Photo Researchers, Inc.)

Materials that can remember their shape have plenty of uses in a variety of situations, whether implanted in the body or orbiting in space. An important potential role in the development of adjustable airplanes will be described in a later section.

Other materials have been found that have shape memories. Shape-memory polymers are plastics that have properties similar to SMAs. Pliable materials have recently been found that can be stretched up to twice their normal length, yet regain their shape when heated. Having a smart plastic material opens up applications of a softer, more flexible nature, such as clothing.

SMART FIBERS—AND SMART CLOTHES

Clothes have long been made of natural fibers such as cotton, wool, and silk. But beginning in the 20th century, textile manufactures have had other options. Nylon was the first synthetic (human-made) fiber, introduced in 1939. Others, such as polyester and spandex, followed. These materials are made of long chains of bonded molecules that form durable fibers. Although natural fibers continue to be used, many clothes today are made from synthetic materials.

If "smart fibers" began to be widely used, perhaps clothes and other fabrics would become more adaptable. Nature is one place to look for inspiration. Olivier Emile and Albert Le Floch of the Université de Rennes in France and Fritz Vollrath at Oxford University in England studied why spiders rarely spin around when hanging from their silk threads. Most fibers turn and twist, as a climber dangling from a rope knows all too well, yet a suspended spider is stable. The researchers discovered that spider silk has a kind of shape memory in which it rapidly recovers its shape, resisting any twisting motion. This research, "Shape Memory in Spider Draglines," was published in *Nature* in 2006.

Researchers who study polymers are looking for novel ways to incorporate shape memory into these materials. Shape-memory polymers that respond to heat have been discovered, and in 2005, Andreas Lendlein at the Institute of Polymer Research in Teltow, Germany, Massachusetts Institute of Technology researcher Robert Langer, and their colleagues found a shape-memory polymer that is responsive to light. As reported in "Light-Induced Shape-Memory Polymers," published in *Nature* in 2005, this smart polymer is composed of certain chemical groups that are sensitive to light. Exposure to light induces the polymer's molecules to bond, cross-linking and stabilizing the structure in a specific shape. Light of different wavelength can break the bonds, returning the polymer to its original shape.

Smart fibers can make smart sutures—threads used by surgeons to sew up incisions—such as suture that ties itself when heated! And Sensatex Incorporated makes a smart shirt that automatically monitors the wearer's vital signs. Special sensors and conducting fibers embedded in the shirt measure heart rate, respiration, and body temperature, and a tiny controller relays this information by transmitter to a medical station.

ADAPTABLE AIRPLANES

One of the most heavily researched applications for smart materials involves aviation. The marvelous adaptability of the wings of peregrine falcons means that the bird is proficient in both soaring and diving.

Soaring bird   Diving bird

© Infobase Publishing

Compare the wing shape of a soaring bird (left) with a diving bird (right).

Although modern airplanes have features such as flaps and rudders that provide control and some degree of adaptability, the change in shape of the wings is modest compared to the ability of peregrine falcons and other birds. The figure above compares the shape of a falcon's wings when soaring and diving. Adaptations as effective as these shape-changing wings are main goals of smart airplane technology.

The U.S. Air Force is particularly interested in developing this technology. Air Force planes must accomplish a variety of different tasks, from ground support to high-altitude bombing. Since airplanes are not very adaptable, each type must be optimized for a single task, which requires specific geometrical and structural properties. The Air Force has heavy bombers to drop high-explosive weapons, maneuverable fighters to engage enemy aircraft, and soaring reconnaissance airplanes to scout enemy positions. Procuring and maintaining a fleet of all these different airplanes is a tremendously expensive job. Any one of these sophisticated airplanes requires millions or even billions of dollars to design and test, and they can be equally costly to build—a B-2 Spirit, for example, has a price tag of more than $2 billion.

Little wonder that the Defense Advanced Research Projects Agency (DARPA), the main research and development agency of the U.S. Department of Defense, initiated a program in 2002 called Morphing Aircraft Structures. The goal of this program is to develop innovative technology that may one day lead to airplanes with wings and possibly other components that can change shape, a process known as morphing. DARPA, which has sponsored a large number of projects over the years that have played an important role in civilian as well as military technology, is discussed further in the sidebar on pages 126–127.

Morphing aircraft could have a large impact on society. One important civilian motivation for morphing aircraft is an increase in efficiency. According to the Air Transport Association, airlines of the United States in 2006 consumed 19.6 billion gallons (74.2 billion L) of jet fuel at a cost of $38.5 billion. Even a minor improvement in efficiency—3 percent, for example—would save a billion dollars, in addition to reducing pollution and other environmental benefits gained by using less fuel.

A model undergoing aerodynamic tests in a wind tunnel *(U.S. Air Force)*

Defense Advanced Research Projects Agency (DARPA)

This agency got its start as the Advanced Research Projects Agency (ARPA), established by a Department of Defense directive on February 7, 1958. The motivation for its establishment is quite clear. On October 4, 1957, the Soviet Union shocked the world when it initiated the space age by launching *Sputnik,* the first artificial satellite. (In Russian, the term *sputnik* means traveling companion.) The United States, the world's leading democracy, and the Soviet Union, then the world's leading communist country, were at the time locked in a fierce political battle of one-upmanship known as the cold war. Fearing that the Soviet Union's science and engineering were outpacing everyone else, the United States decided to close the gap by establishing research agencies and increasing the amount of money spent on research.

On March 23, 1972, ARPA's name changed to Defense Advanced Research Projects Agency (DARPA). (Then-president Clinton's administration switched the name back to ARPA on

Some types of airplane used on board Navy aircraft carriers can fold their wings. But these wings are heavy and inefficient—they are designed to fold in order to save space in the severely cramped quarters of a ship, rather than any attempt to increase flight capability. The goal of morphing projects is to design and build wings and other structures that could change shape in flight. In order for the airplane to excel at different tasks, a 50–75 percent change in wing area will probably be necessary.

Smart systems involve sensing changing conditions and activating some sort of response, usually with an actuator. A project known as Active Aeroelastic Wing (AAW), conducted by the National Aeronautics and Space Administration (NASA), the U.S. Air Force Research Laboratory, and Boeing Company, studied wing elasticity in a modi-

February 22, 1993, but three years later the title reverted to DARPA.) Based in Arlington, Virginia, DARPA has a staff of about 250 people and a budget of several billion dollars. Its approach is to develop and fund projects that target specific near-term or future technologies such as morphing aircraft, high-energy lasers, chemical sensors, energy alternatives such as biofuels, and many others. Individual projects last some number of years. DARPA staff determine which technologies to target, then set reasonable goals and fund research conducted by private research organizations or university laboratories.

DARPA projects have led to the development of stealth technology that allows airplanes such as the B-2 and F-117 to elude radar detection, as well as the unmanned, remotely piloted aircraft that have plagued terrorists in the Iraqi conflict, and other technological advances. But DARPA projects have also played important roles outside of the military. The most popular and enduring such project began in the 1960s with a network of computers, known as ARPANet, which permitted users to send and receive electronic messages and share programs and other items. This network eventually became the Internet.

fied F/A-18A. Lighter, more flexible wings have been tested, along with actuators and controls by which the outer wing panels twisted up to an angle of five degrees. Flight tests carried out in 2005 proved that the concept worked—controlled flexibility of the lightweight "warping" wings provided stable flight even at supersonic speeds. Thanks to Newton's second law of physics—acceleration of an object equals force divided by the object's mass—a lighter, less massive airplane is more efficient since it experiences a higher acceleration for a given force. In a report on the project after the 2005 flight tests, NASA was optimistic: "With the successful demonstration of actively controlled 'wing warping' techniques for aircraft roll control at transonic speeds [close to the speed of sound] in the Active Aeroelastic Wing project, engineers will now have more freedom in designing more efficient, thinner, higher aspect-ratio wings for future high-performance aircraft while reducing the structural weight of the wings by 10 to 20 percent."

Research teams working on DARPA's Morphing Aircraft Structures project are even more ambitious. Cornerstone Research Group, Inc. (CRG) and Lockheed Martin are testing a wing capable of withstanding changes of up to twice the length. The wing's skin is made of a shape-memory polymer designed by CRG scientists and engineers. When heated by flexible wires hidden in the structure, the material softens in a few seconds and morphs into a new shape. As it cools, the polymer sets, keeping its shape until another change is needed.

FlexSys, Inc., funded by a U.S. Air Force program called Small Business Innovation Research, has made and tested a wing with a trailing edge that can flex up and down as much as 10 degrees. The adaptive system consists of aluminum and composites—standard aeronautic materials—driven by actuators. Flight tests began in 2006, using an airplane called the White Knight, which was built by Scaled Composites, Inc. (White Knight is the airplane from which SpaceShipOne launched in 2004—SpaceShipOne was the first private vehicle in space.) The wing is not actually used for lift at this early testing stage; instead, a small version of the wing, measuring 30×50 inches (75×125 cm), is attached to White Knight's underside. FlexSys researchers are studying the wing's behavior as indicated by gauges and sensors connected to the surface. Further testing will be needed to determine if a full-scale version can lift and control an airplane in flight.

A number of important issues remain unresolved. Drastic changes during flight raise a number of problems. Highly adaptable structures such as morphing wings may be perfectly stable and controllable while set in one shape or another, but the transition between shapes might cause trouble in flight. As an analogy, imagine two people in a narrow canoe. The craft is stable in the water as long as one person stays in the front and one person in the rear. But there is a problem if they want to switch seats—while passing each other in the middle, the boat is liable to tip over. For an airplane in flight, instability during a transition can have consequences a lot more tragic than getting wet.

But flight stability with morphing wings is possible—consider the peregrine falcon. Swifts are another kind of acrobat flier. Researchers in the Netherlands and Sweden have studied these birds in flight and have even tested their wings in a wind tunnel. The result, "How Swifts Control Their Glide Performance with Morphing Wings," published by David Lentink and his colleagues in *Nature* in 2007, showed that by changing the shape of their wings, swifts can fly up to 60 percent farther and turn up to three times faster.

CONCLUSION

Smart materials and structures often mimic living organisms, which are the ultimate smart systems. Inspired by highly adaptable animals such as falcons and swifts, smart materials are becoming increasingly important in how chemists, engineers, and materials scientists think about and design objects. As in clothes, one size does not fit all. A smart material that changes to meet varying conditions is much more efficient than making a different system or structure for all occasions.

Modified F/A-18A used for the Active Aeroelastic Wing project (NASA, Jim Ross)

How far can smart materials go? Futuristic depictions such as in the 1991 film *Terminator 2* have androids made of shape-changing metal that can quickly flow and set into any desired form. A more realistic vision for the future of smart materials and systems can be viewed by observing nature. Organisms move, adapt, and evolve, and they are made of materials that are complex but have been studied by biologists for decades.

The human body, for instance, has sensors (eyes, ears, touch receptors in the skin, and so forth), a controller (the brain), and actuators (muscles) to react and respond to commands. These are the same basic concepts as the adaptive systems discussed in this chapter. Robots today, such as the welding machines used in industry or the toy dogs sold as pets, are extremely limited in mobility and adaptability compared to humans. Yet smart materials, along with a design based on the sensory, nervous, and muscular systems of the body, could one day create an agile and adaptable robot.

Researchers are not even close to producing such a robot, but progress toward distant goals often proceeds a step or two at a time. Peter Bentley, a researcher at University College London in England, along with his student, Siavash Haroun Mahdavi, designed and built a robot snake in 2003 that had "muscles" made of nitinol wire. The controller was not given a specific set of instructions but instead was allowed to evolve and create its own method of locomotion, which it did—and the motion resembled the wavy movements of a cobra. When one of the nitinol wires broke, the system changed to its new situation and evolved a slightly different method of locomotion that mimicked an earthworm.

Adaptation and evolution have created an astonishing variety of life on this planet. Smart materials and adaptive systems may do the same for instruments, machines, and other technological tools and techniques. The expansion of this scientific frontier has the potential to revolutionize much of the technology that people use for medicine, transportation, and industry.

CHRONOLOGY

1824 c.e.   Although similar effects had been seen and written about before, Scottish scientist Sir David Brewster

(1781–1868) mathematically describes the alteration in electrical properties of Rochelle salt with a change in temperature.

1842   British researcher James Joule (1818–89) discovers the magnetostrictive effect in a sample of iron that changes length in response to a magnetic field.

1880   Frenchman Jacques Curie (1856–1941) and his younger brother Pierre (1859–1906) discover piezoelectricity. (Pierre Curie would go on to make many other discoveries and marry Marie, a Polish scientist, who also became famous.)

1903   American inventors Orville Wright (1871–1948) and Wilbur Wright (1867–1912) make the first sustained flight in North Carolina with an airplane controlled by warping the wings.

1932   Swedish researcher Arne Olander observes a gold-cadmium alloy that can be bent while cool but returns to its original shape after heating—a shape-memory alloy.

1947   Willis M. Winslow, an engineer and inventor, files a U.S. patent that includes the first description of an electrorheological fluid.

1948   Jacob Rabinow, a researcher at the National Bureau of Standards (since renamed as the National Institute of Standards and Technology), discovers the magnetorheological effect.

1963   William J. Buehler and his colleagues at the Naval Ordnance Laboratory in White Oak, Maryland, describe the shape-memory alloy known as nitinol.

1965   William H. Armistead and Stanley D. Stookey, researchers at Corning Glass Works in New York, receive a patent for "phototropic material," an early photochromic glass.

1967	The U.S. Air Force jet F-111, a pioneering "swing wing" (adjustable wing position) aircraft, becomes operational.
1980s	Robert Q. Fugate, a researcher working for the U.S. Air Force, and his colleagues and other researchers develop adaptive optics.
1999	Keck Observatory on Mauna Kea, Hawaii, installs an adaptive optics system.
2002	Defense Advanced Research Projects Agency (DARPA) begins a project, Morphing Aircraft Structures, to investigate the potential of such technology as shape-changing wings.
2005	National Aeronautics and Space Administration successfully tests the Active Aeroelastic Wing.
2006	Gregory Carman, Lenka Stepan, and Daniel Levi design and build a heart valve made with nitinol.

FURTHER RESOURCES
Print and Internet

Beylerian, George M., Michele Caniato, Andrew Dent, and Bradley Quinn. *Ultra Materials: How Materials Innovation Is Changing the World.* New York: Thames & Hudson, 2007. This well-illustrated book focuses on smart materials developed for applications in textiles, fashion, and design.

Center for Adaptive Optics. "Adaptive Optics." Available online. URL: http://cfao.ucolick.org/ao/. Accessed May 28, 2009. This tutorial on adaptive optics has sections explaining the principles of adaptive optics, why it is useful, and how it works.

Cornerstone Research Group. "Shape Memory Polymers—An Overview." Available online. URL: http://www.crgrp.net/overviews/smp1. shtml. Accessed May 28, 2009. Makers of shape memory polymers, Cornerstone Research Group explains and illustrates the concepts.

Dick, Ron, and Dan Patterson. *Aviation Century: Wings of Change.* Ontario, Canada: Boston Mills Press, 2005. Many photographs enhance this volume's description of the progress in aviation in the era following World War II. The final chapter discusses aeronautical research.

Eberhart, Mark. *Why Things Break.* New York: Harmony Books, 2003. Although it is not about smart materials, this book describes the properties of materials that researchers hope to improve with innovative technology such as smart materials. Plenty of examples are given of what can go wrong in the engineering design and manufacturing of vehicles and structures.

Emile, Olivier, Albert Le Floch, and Fritz Vollrath. "Shape Memory in Spider Draglines." *Nature* 440 (March 30, 2006): 621. The researchers discovered that spider silk has a kind of memory in which it rapidly recovers its shape.

EyeWitness to History.com. "The Wright Brothers—First Flight, 1903." Available online. URL: http://www.eyewitnesstohistory.com/wright. htm. Accessed May 28, 2009. The Wright Brothers first flight is described, including a long passage from Orville Wright's diary.

FlexSys, Inc. "Mission Adaptive Compliant Wing." Available online. URL: http://www.flxsys.com/Projects/MACW/. Accessed May 28, 2009. The company provides information, pictures, and results on its flexible wing project.

General Motors. "Intelligent Chassis Control Systems: Taking Safety Along for the Ride." 2003. Available online. URL: http://media.gm.com/ division/2003_prodinfo/03_corporate/chassis.html. Accessed May 28, 2009. General Motors describes suspension systems that incorporate smart materials.

Lendlein, Andreas, Hongyan Jiang, Oliver Jünger, and Robert Langer. "Light-Induced Shape-Memory Polymers." *Nature* 434 (April 14, 2005): 879–882. The researchers report the discovery of a shape-memory polymer that is responsive to light.

Lentink, D., U. K. Müller, E. J. Stamhuis, R. de Kat, W. van Gestel, L. L. M. Veldhuis, et al. "How Swifts Control Their Glide Performance with Morphing Wings." *Nature* 446 (April 26, 2007): 1,082–1,085. The researchers conducted wind tunnel experiments demonstrating that by changing the shape of their wings, swifts can fly up to 60 percent farther and turn up to three times faster.

Musolff, André. "Shape Memory Alloys." Available online. URL: http://www.smaterial.com/SMA/sma.html. Accessed May 28, 2009. This richly illustrated and highly informative Web site describes shape-memory alloy from the perspective of models, crystallography, simulation, applications, and research.

National Aeronautics and Space Administration. "Active Aeroelastic Wing Flight Research." Available online. URL: http://ims.ivv.nasa.gov/centers/dryden/pdf/120314main_FS-061-DFRC.pdf. Accessed May 28, 2009. This NASA report describes the results of the Active Aeroelastic Wing project.

University of Alberta. "Educational Software for Micromachines and Related Technologies." Available online. URL: http://www.cs.ualberta.ca/~database/MEMS/sma_mems/index2.html. Accessed May 28, 2009. Research groups at the University of Alberta in Canada constructed this Web resource, which discusses a variety of smart materials, including shape-memory alloys, piezoelectric materials, and electrorheological and magnetorheological fluids.

Web Sites

Defense Advanced Research Projects Agency. Available online. URL: http://www.darpa.mil/. Accessed May 28, 2009. DARPA's Web site explains its mission and goals and contains a news section describing the latest research, along with historical information, located on the "DARPA Legacy" page, about previous projects and results.

National Aeronautics and Space Administration—Ames Education Division: Smart Materials. Available online. URL: http://virtualskies.arc.nasa.gov/research/youDecide/smartMaterials.html. Accessed May 28, 2009. As part of an educational activity in which students plan an aviation research project, this Web site provides links to pages discussing piezoelectric materials, electrorheological and magnetorheological fluids, shape-memory alloys, and magnetostrictive materials.

5

FUEL CELLS—
ENERGY FOR A
POWER-HUNGRY
WORLD

In 1903, when Wilbur and Orville Wright made the first sustained flight in a powered aircraft, a gasoline engine was the best choice to supply the power. Alternatives to gasoline engines were available at the time but had disadvantages—a steam engine would have been too heavy, and an electric motor might not have produced sufficient power. The four-cylinder gasoline engine the Wright brothers built generated 12 horsepower, which is not much more powerful than one of today's riding lawn mowers, but it managed to keep the airplane off the ground for its short flight. Five years later, in 1908, Henry Ford's first Model T used a gasoline engine, and the popularity of this car resulted in widespread usage of gasoline engines.

Today, there are nearly a billion cars on the world's roads. Most of them run on gasoline or diesel fuel. According to the Department of Energy, the U.S. government agency that monitors and researches energy sources, the United States alone consumes about 140 billion gallons (532 billion L) a year, and U.S. airlines burn about 20 billion gallons (76 billion L) of jet fuel annually. Burning an energy-rich fuel is efficient and, despite recent price increases, still reasonably cheap—although a gallon (3.8 L) of gasoline cost only about 30 cents in 1920, the price is not much higher these days when adjusted for inflation.

But Earth and its resources are not boundless. Gasoline is made from petroleum—oil pumped out of the ground. (Gasoline should not be confused with natural gas, which is mostly methane and is burned to heat homes, make electricity in generators, and for other purposes.) The planet does not have an infinite supply, and although no one is sure how long the supply of petroleum might last, high consumption rates may exhaust this resource within 50–100 years.

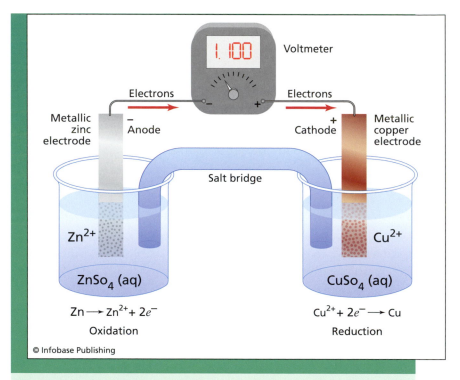

Voltmeter

Electrons   Electrons

Metallic
zinc   − Anode   + Cathode   Metallic
electrode   copper
electrode

Salt bridge

Zn^{2+} Cu^{2+}

$ZnSo_4$ (aq) $CuSo_4$ (aq)

$Zn \longrightarrow Zn^{2+} + 2e^-$ $Cu^{2+} + 2e^- \longrightarrow Cu$

Oxidation   Reduction

© Infobase Publishing

The left half of this electrochemical cell, containing a zinc electrode and zinc sulfate solution, engages in an oxidation reaction. This reaction liberates electrons and turns zinc atoms into zinc ions (Zn^{2+}, which is a zinc atom that has lost two electrons and therefore has a net positive charge of 2). The zinc atoms come from the electrode, which is gradually depleted; the zinc ions that are produced in the reaction enter the solution. In the right half of the cell, a reduction reaction occurs; electrons combine with copper ions to produce neutral atoms of copper. Copper ions leave the solution in the process and collect at the copper electrode. Over time, the zinc electrode and copper solution will run out of material, causing the reaction to cease unless the material is replenished.

Another problem is also looming. Earth is warming—surface temperatures have increased about 1.33°F (0.745°C) in the last century. Scientists of the Intergovernmental Panel on Climate Change issued a report in 2007 and concluded, "Most of the observed increase in global average temperatures since the mid-20th century is *very likely* due to the observed increase in greenhouse gas concentrations." These atmospheric greenhouse gases, such as carbon dioxide, retain the planet's warmth, similar to actions of a greenhouse, and elevates the temperature. (The term *gas* as used here refers to a phase of matter, as in air.) This rise in temperature may not seem like much, but the warming trend has been melting polar ice. If the melting continues unabated, an alarming rise in sea level may ensue, along with other drastic climate changes that occur when nature's delicate balance is upset. Combustion of gasoline and similar fuels produces greenhouse gases such as carbon dioxide. Yet Earth's burgeoning population and the essential economic development to support this population require increasing amounts of energy. To meet these needs while maintaining the integrity of the environment, alternative sources of energy must be found. One alternative is an electrochemical device known as a fuel cell, the subject of this chapter.

INTRODUCTION

Electrochemistry is the study of chemical reactions in which electricity plays a role. Some electrochemical reactions generate electricity as the reaction proceeds, while in other cases the opposite occurs—electricity drives the reaction. In either case, electrochemical reactions involve the transfer of electrons, which are the negatively charged particles surrounding an atom's nucleus. Reactions in which electrons are transferred (or appear to be transferred) from atom to atom are called *oxidation-reduction* reactions.

An oxidation reaction is so named because it used to refer to chemical activities involving oxygen, such as combustion. The term *combustion* is presently used in a broader sense, and it describes a process by which an element increases its oxidation number. An oxidation number is an abstraction—chemists assign this number based on a set of rules, which helps them understand reactions. Oxidation corresponds to a loss of electrons. Reduction, on the other hand, corresponds to a gain in electrons. Miners and metal producers have long used reduction reactions, for example, when they chemically treat certain metal ores to extract the metal—the result is a purer metal, which reduces the mass of the original ore.

The transfer of electrons sets up an electric current—a flow of electric charges. This flow is vital in electrochemical reactions. As the reaction continues, electrons are injected into the process and then drawn off. Chemists carry out electrochemical reactions within a device known as an electrochemical cell, with electrical conductors called *electrodes* to inject and withdraw electrons.

An electrochemical cell should not be confused with a biological cell, which is the basic unit of life. Electrochemical cells contain chemical solutions and electrodes to conduct electrons. As shown in the figure on page 136, both oxidation and reduction reactions occur in the cell, but these reactions are separated. Separation is essential so that electrons can flow through electrodes and into attached wires, which can be routed to wherever electricity is needed. To maintain electrical balance at the electrodes—in other words, to complete the circuit so that electricity flows in a loop—an *electrolyte* allows the flow of ions between the two halves of the system. This cell produces electricity and is called a voltaic cell. The reaction proceeds as long as the materials last.

In contrast to a voltaic cell, electrolysis is a type of electrochemical reaction in which electricity is the driving force rather than the product. These reactions occur only with an energy input, as provided by the electric current. In electrolytic cells, the electrodes would be attached to some sort of electric generator to push electrons through the cell. For example, electrolysis of water (H_2O) is a reaction in which the components, hydrogen and oxygen, are released. (*Lysis* is a term derived from a Greek word meaning to loosen or break apart. Electrolysis is the process by which electricity breaks apart a compound such as water.)

Electrodes in a voltaic cell, however, are connected to circuits— paths by which electrons flow. Voltaic cells are sources of electricity, so they can be used to drive electrolytic reactions or perform other activities that require electricity. The term *voltaic* honors the Italian scientist Alessandro Volta (1745–1827), a pioneer of electrochemistry. A simple voltaic cell can form a battery, invented by Volta in 1800. The unit of electric potential, the volt, also honors Volta.

Batteries are needed to power small, portable instruments such as flashlights, cameras, and computers. In many flashlight batteries, the two electrodes—attached to the plus and minus ends of the battery—

consist of a rod of carbon in the center, and the casing, which is made of zinc. An electrolyte composed of a paste of zinc chloride or similar compound carries the ions. The electrons flow from the negative terminal, which is the zinc case, through whatever electrical circuit is attached to the battery, and then back to the battery's positive terminal—a metal cap on the carbon rod.

A battery such as the zinc-carbon battery produces a small amount of electricity, the quantity and duration of which depends on the chemical reaction. The reaction eventually depletes the materials, and when the reaction stops, the battery is "dead." Some batteries are rechargeable—passing an electric current though the battery causes the reaction to run in reverse, restoring the original materials and allowing the battery to work again. But inexpensive batteries such as zinc-carbon batteries are not rechargeable and are discarded.

Electricity is so prevalent in modern society that an electrical power failure, such as the "blackout" in the northeastern part of the United States and parts of Canada on August 14, 2003, causes a major inconvenience. People rely on electricity to heat and light their homes, to operate many different kinds of appliances and motors, and to run subways, commuter trains and similar transportation systems. There is also a great deal of reliance on batteries, which store electricity and allow the portability of small devices that would otherwise require the use of wires to carry the required current from a generator.

Although in principle batteries can power any device that runs on electricity, in many cases the amount of electricity would require an excessive number of batteries. Another problem is that batteries have a limited life or, in the case of rechargeable batteries, require frequent recharging. But imagine an electrochemical system in which the reactants continually flowed in the cell. The consumed reactants are replaced as the reaction proceeds, so the cell can function continuously with no need for recharging. In this situation, the flowing reactants can be considered as the "fuel," and the cell is known as a fuel cell.

FUEL CELLS

Ever since Volta's first battery in 1800, scientists have been trying to develop bigger and better electrochemical cells. Sir William Robert Grove (1811–96), a lawyer from Wales who had a penchant for experimental science, discovered an interesting cell in 1838. He placed one end of a set of platinum electrodes in an electrolyte and the other end in a container of oxygen or hydrogen gas. When he connected the electrodes with a wire, an electric current flowed. As the current flowed, Grove observed water accumulating in the gas containers. Grove was puzzled, but at about the same time Grove published his experiment on "gas batteries," the Swiss chemist Christian Schönbein (1799–1868) published the first description of a fuel cell.

Electrolysis of water, mentioned above, had been described by the British chemists William Nicholson (1753–1815) and Sir Anthony Carlisle (1768–1842) in 1800. But Grove's experiment seemed to go in the opposite direction. This "reverse eleoctrolysis" is the basic operation of the fuel cell—the combination of hydrogen gas (H_2) and oxygen gas (O_2) to produce water and energy, as described in the following chemical equation:

$$2 H_2 + O_2 \rightarrow 2 H_2O + \text{energy}$$

Hydrogen, discovered by the British chemist Henry Cavendish (1731–1810) in 1766, is well known to enter this reaction, as implied by the element's name. *Hydro* refers to water, and *gen* refers to generation—hydrogen generates water when burned in air. Grove and Schönbein showed how this reaction can be used to create energy in the form of electricity, although in this case the "burning" is slow and controlled.

The figure opposite depicts a typical hydrogen fuel cell. On the left side, called the anode, hydrogen or some gas containing hydrogen is pumped into the chamber. The electrode, commonly made of platinum, catalyzes the oxidation of hydrogen, which produces hydrogen nuclei (protons) and electrons. The protons reach the cathode (positive) chamber by way of an electrolyte, but the electrons travel through a conductor. Catalyzed by another platinum electrode, oxygen pumped into the cathode side reacts with protons and electrons to produce water. The electrons flowing through the conductor form a current, and the fuel cell continuously supplies electricity.

Not just any electrode will do. Catalytic electrodes are essential because they speed up the reactions so that the process creates a sizable current. A similar situation occurs in biology, where catalysts known as enzymes increase the rate of vital biochemical reactions that would otherwise be too slow to allow the survival of living organisms. Catalysts participate in the reactions but are not consumed by them.

Many different types of fuel cell are possible. The fuel in the cell described above is pure hydrogen, although of course an oxidizing agent, in this case oxygen, is also needed. Other fuels can be used, although the reactions are more complicated. Various electrolytes are also possible. For example, the electrolyte of alkali fuel cells is a solution of potassium hydroxide (an alkaline, which has a high pH). Solid oxide fuel cells use compounds of metal oxides as electrolytes. (Fuel cell names are generally

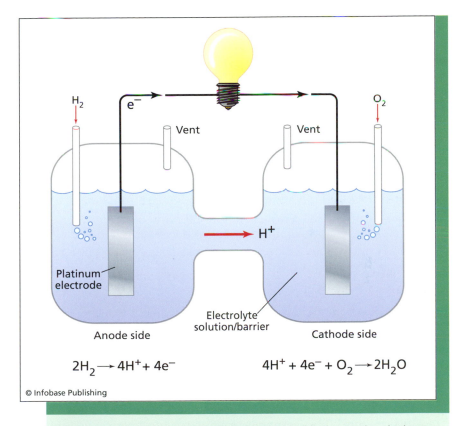

$$2H_2 \longrightarrow 4H^+ + 4e^-$$

$$4H^+ + 4e^- + O_2 \longrightarrow 2H_2O$$

© Infobase Publishing

Platinum electrodes catalyze the breakdown of hydrogen into hydrogen nuclei (protons) and electrons. The electrons travel in a conductor, and the protons travel through the electrolyte, meeting at the other side, where platinum electrodes catalyze the combination of protons, electrons, and oxygen into water.

Series and Parallel Circuits

Imagine a row of batteries with the positive end of one battery against the negative end of the next battery. Part A of the figure opposite illustrates this situation, which is called a series circuit. One way of thinking about how a circuit works is to think of the flowing charges as "falling" from an upper level to a lower one. As the charges "fall," they move around the circuit, constituting a current. Although a charge falls naturally from a high level to a low one, it will not climb back to the high level without some help. A battery is a device that lifts an electric charge from a lower level, or voltage, to a higher one. Batteries are needed to boost the charge back up to the higher level—that is, to generate a voltage—so that charges can continue to flow around the circuit. In a series circuit, a charge encounters the row of batteries, each of which gives it a small "elevation." A series of batteries therefore "raises" a charge to the sum of their voltages.

Batteries in parallel, as diagrammed in Part B, have their own separate path. A charge flowing in this circuit encounters only one of the batteries. The charge receives only a little "lift"—the voltage applied by a single battery (all of which should have the same voltage). But because of the separate paths, the parallel configuration can push a lot more charges than a series configuration, though it is somewhat risky—if one of the batteries in a parallel configuration fails, it could create a "short" that draws all the current, resulting in a fire hazard.

In other words, voltages in series add, but voltages in parallel do not. This principle holds true for batteries, fuel cells, and any other source of electricity. High voltages are sometimes required to give charges an extra push, such as when the charge must pass through an area of high resistance. To meet this need, a high-voltage source might be used, or some number of low-voltage batteries can be placed in series.

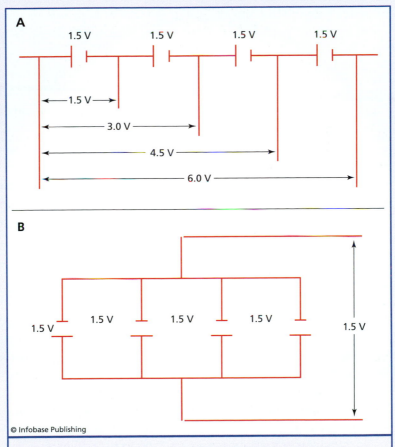

(A) The voltages of batteries in a series. The same current flows through each battery in the circuit. (B) The voltages of batteries in parallel do not add, because a charge flows through only one of the circuit branches.

based on the name of the electrolyte.) Such variety allows fuel cells to operate in a number of different conditions and temperatures.

As in batteries, fuel cells generate an electric potential, measured in volts, which depends on the reaction. One of the simplest fuel cells is called proton exchange membrane—a permeable membrane acts as an electrolyte by conducting positive charges such as protons but not electrons. This fuel cell generates about 0.7 volts, which is about half the voltage of a common flashlight battery. Small voltages are sufficient to power only the tiniest devices. But the lack of high voltage does not spoil the use of fuel cells since the voltage can be increased by making a fuel cell stack. A stack of fuel cells puts a number of fuel cells in a row, creating a series circuit. As described in the sidebar on pages 142–143, the voltages of batteries in series add or combine to produce a higher voltage. To create a stronger current, fuel cells should be placed in parallel.

Although Grove's cell worked, it was inefficient and not practical for everyday use. The earliest practical fuel cells did not appear until the 1950s. In 1959, the British engineer Francis Thomas Bacon demonstrated an alkaline fuel cell capable of generating enough power to operate a welding machine. Around the same time, Willard Thomas Grubb, a chemist working at General Electric, designed a fuel cell using a membrane. These early fuel cells were expensive, but they soon filled a need for which other available electricity sources were not well suited—going beyond the planet.

POWERING SPACECRAFT

On October 4, 1957, the Soviet Union, as it was then known, launched *Sputnik,* the world's first artificial satellite. This feat of technology not only inaugurated the space age, it also spurred the United States into investing a lot of time and money in its own space program. The plans included an ambitious series of manned spaceflights, beginning in the 1960s, eventually leading to a landing on the Moon in 1969.

Spacecraft need power. Rockets provide propulsion, but the equipment needed for control of the spacecraft and to provide life support for the astronauts also need a source of energy. Much of this equipment runs on electricity. One possible solution was to use batteries, but the National Aeronautics and Space Administration (NASA), the U.S. agency responsible for space exploration, rejected this idea. The problem with batteries was that the power needs of manned space flight would require a large number, especially since safety concerns necessitated carrying plenty of backups—once the astronauts left Earth, they had to have everything they would ever need already on board. Weight is a serious issue in rocket launches; escaping Earth's gravity requires tremendous acceleration, and because of Newton's second law of motion—acceleration equals force divided by mass—increasing the mass of a rocket increases the force required to accelerate it. NASA engineers were worried that even their most powerful rocket would be unable to accelerate safely to the required speed if the mass was too great. Every item was weighed, and engineers excluded anything not essential to the mission.

Space shuttle launch *(NASA)*

Another power option was to derive energy from sunlight—solar energy—but in the 1960s, the technology to do this was not quite practical. Nuclear energy techniques were available, but NASA felt this would be too risky as well as too heavy.

The best alternative proved to be fuel cells. NASA chose an alkaline fuel cell, similar to the one Bacon had built, for the Apollo missions, including *Apollo 11,* which landed on the Moon on July 20, 1969. These fuel cells were reliable, efficient, and performed another valuable service in addition to providing electricity—the reaction produced drinking water for the astronauts. Three fuel cells, each capable of 28 volts and weighing 250 pounds (114 kg) on Earth's surface, operated in parallel on the Apollo flights. Just one of these fuel cells would ensure a successful mission, so two extra cells gave the astronauts a comfortable safety margin. But the fuel cells performed extremely well, logging thousands of hours of operation without a failure.

Fuel cells have also been used for *Skylab,* a space station launched on May 14, 1973, and in orbit for more than six years, as well as in the space shuttles. Other uses for these reliable devices include providing electricity at remote facilities that are beyond the reach of power lines, and as back-up generators at hospitals, which cannot afford to lose power for any length of time.

There is also the possibility of propelling vehicles. This means not just providing electricity, as with NASA spacecraft, but also providing the means of propulsion. Space-faring rockets require a bit too much power for this to be practical as yet, but cars and small airplanes travel at much more attainable speeds. Fuel cell engines are an extremely active area of research.

Several important factors motivate this research on fuel cell engines. One of these factors is the efficiency of fuel cells when compared to engines that burn gasoline, as discussed in the following section.

ENGINES AND THERMODYNAMICS

The Wright brothers and Henry Ford selected a gasoline engine because it was the best bet at the time, and it remains a popular and commonly used engine today. But even in the early 20th century, scientists knew that this type of engine has limits. An engine that burns gasoline or similar fuel is a type of heat engine, which is a category of engines that convert heat into work. The basic idea is to generate a hot, expanding gas; in gasoline engines, this is accomplished by igniting gasoline, while older steam engines accomplished the same goal by burning coal or wood to boil water and make steam. Expanding gases push against a movable rod or some other object, which in turn may rotate an axle to power a car or drive a turbine to power a jet airplane.

Heat engines are relatively easy and inexpensive to build. Yet not all the heat can be converted into work. Imagine lighting a fire in the fireplace to warm a room on a cold winter day—the temperature of the room increases, but a lot of energy is wasted because some of it heats air that rises and escapes up the chimney. Similar losses occur in heat engines. Engines get hot, and their temperature increases so much that they must be cooled or they will overheat and the parts will melt. This heat is wasted energy because it is not converted into work such as moving a car or turning a pump. Efficiency is a measure of how much of the available energy is converted into work, or in other words, how little is wasted. For example, an engine with 50 percent efficiency converts half of the available energy into work.

Engineers can reduce some of the heat loss by careful design, but can a heat engine ever be designed to prevent all heat loss? This is a question addressed by thermodynamics, the study of heat and energy. (The word *thermodynamics* comes from a Greek term, *thermē,* meaning "heat.") The answer, much to the dismay of engineers, is no. No heat engine, no matter how it is designed or what kind of fuel it burns, can ever be 100 percent efficient.

This limitation, imposed by a scientific law called the second law of thermodynamics, can be difficult to understand. It involves a concept known as entropy, which can be thought of as a measure of disorder. Entropy must increase in natural processes; in other words, processes naturally go from order to disorder (as observed by anyone who has bought a shiny new bicycle or automobile and watched it fade, corrode, break down, and finally fall apart—usually just after the warranty expires). The second law of thermodynamics requires a heat engine to vent some heat into the environment, thereby raising entropy. This loss is unavoidable, and a heat engine will not operate without it. No one will ever buy a car powered by a gasoline engine that does not exhaust, and lose, some of its heat.

The good news for fuel cells is that they run on a different process. Fuel cells are not exempt from scientific laws, but their manner of energy conversion is electrochemical rather than thermal. The maximum efficiency for the electrochemical processes in fuel cells is higher than for the internal combustion engines that power many automobiles today.

But to reach these maximum efficiencies, researchers and engineers must find the optimal design. Even the most modern internal combustion engine does not operate at its highest possible efficiency—additional heat is lost through the parts of the engine, for example, which is not required by thermodynamics but occurs because engineers cannot find a suitable material that will prevent the loss. (Such a material should be a thermal insulator, which would prevent heat from escaping. The problem is that engines require material that is strong and able to withstand high temperatures, and material having these properties tend to be metals or alloys—which are thermal conductors, not insulators.) Fuel cells also need careful design considerations or they will fail to reach their optimal efficiency.

Yet the advantages of fuel cells over heat engines are not just economical. In addition to heat, gasoline engines emit a lot of pollutants.

A CLEAN SOURCE OF ENERGY

Gasoline consists of a mixture of hydrocarbon molecules, which are long chains of hydrogen and carbon, along with a few other added chemicals. Octane, for example, contains eight carbon atoms and 18 hydrogen atoms and is an important component of gasoline. Burning a substance such as a hydrocarbon is the result of a rapid reaction with oxygen (or some other oxidizing agent), which produces a lot of heat. In a cylinder of an automobile gasoline engine, a spark from a spark plug ignites a mixture of gasoline and air, resulting in an explosive reaction that creates hot, expanding gases. These gases push against a piston, turning the car's crankshaft. The gases are then released into the exhaust to prepare for the next combustion cycle. (Exhaust gases are still relatively hot—the second law of thermodynamics requires this heat loss.) If all of the hydrocarbons in the fuel were burned, the gaseous products would be water, H_2O, and carbon dioxide, CO_2.

The release of carbon dioxide into the environment has come under scrutiny lately because of its possible connection to global climate change. A single car's emission is not a big problem, but the exhaust of millions of cars, along with other combustion engines, adds up.

The emission problem is made worse by the incomplete combustion that occurs in most heat engines. Automobile engines produce a lot of power, which requires a lot of activity in each cylinder (up to 15 or 20 combustions per second, and even more for high-performance vehicles). The cycle happens so quickly that some of the fuel does not burn completely. As a result, the gases emitted in the car's exhaust not only consist of water and carbon dioxide, they also contain unburned hydrocarbons as well as noxious chemicals such as carbon monoxide, which is a poisonous gas. In addition, nitrogen oxides such as nitrogen dioxide (NO_2) are created by the high temperature of the engine and the presence of air, which is about 78 percent nitrogen and 21 percent oxygen by volume. Nitrogen oxides contribute to the stifling smog and haze that is often experienced in large cities.

Polluted air smells bad and makes city dwellers long for fresh air. But air pollution has even more dire effects for both animals and humans. Diseases in humans that are aggravated by air pollution include asthma, emphysema and other lung diseases, and respiratory allergies. The problem had become so bad by 1970 that the U.S. government passed the Clean Air Act. This act gave the Environmental Protection Agency (EPA) authority to establish standards and to regulate vehicle emissions. Over the years, a series of laws have tightened the standards, forcing automobile manufacturers to introduce and improve emission control systems such as catalytic converters, which catalyze reactions that help break down emissions. (A small amount of platinum is often used in these converters, along with other materials.)

Although the EPA and other regulatory agencies have helped reduce the rate of air pollution, the problem persists. Automotive emissions and other pollution sources can be decreased but not eliminated—unless, that is, cleaner sources of energy can be found. The Energy Policy Act of 2005, signed into law by then-president George W. Bush on August 8, 2005, encourages the use and advancement of clean sources of energy by providing tax breaks and other incentives.

One of the cleanest energy sources are fuel cells running on hydrogen. With hydrogen as fuel, emissions are nothing but water—this is considered to be "zero emission" because no harmful chemicals or carbon dioxide are emitted. But this is true only for fuel cells that run on pure hydrogen. Use of hydrocarbons such as methanol for fuel results in other emissions, although the amount is generally less than that generated by combustion engines.

Considering the emissions, why would anyone want to operate a fuel cell on anything but hydrogen? Obtaining pure hydrogen is the problem since there are few natural sources of this element. There are many ways of extracting hydrogen from hydrocarbons and other hydrogen-bearing substances, but these processes can emit considerable pollution. A fuel cell that burns hydrogen is a clean source of energy, but if the method of obtaining the hydrogen is not clean, the overall goal of zero emissions is not achieved.

Finding or producing a clean source of hydrogen is just one of the difficulties fuel cell researchers face. The previous sections of this chapter have discussed why fuel cells are desirable and how they have proven their reliability in the Apollo missions and elsewhere, but numerous problems remain to be solved before this frontier of science can develop cost-effective devices and achieve the promise of a clean, efficient energy source.

FINDING THE BEST CATALYST

One of the most daunting issues is the catalyst. The reactions at the electrodes tend to go slowly, and making or coating the electrode with a catalyst is essential to speed things up. Platinum is one of the most effective catalysts because it binds the reactants and holds them in place so that the reaction can proceed. But the problem is cost—platinum is a rare metal and not at all cheap. An ounce (28.6 g) of platinum costs about $1,200 as of May 2009. Compare that to gold, a precious metal that costs about $970 per ounce (28.6 g) as of May 2009. Fuel cell electrodes would be cheaper if they were made of gold (but gold is not an effective catalyst)!

Fuel cell power is often measured in watts or kilowatts (1,000 watts), as is most electrical devices. Scientists use the term *power* as a measure of the energy per unit time. (A watt equals one joule, an energy unit, divided by one second.) For example, 60 watts is a common power rating for an incandescent lightbulb. In the United States, the power of automobile engines is often described in horsepower, an old unit for power. One horsepower equals about 746 watts. A typical automobile engine is rated at 170 horsepower, or 127 kilowatts. Producing this much power from a fuel cell requires only a few ounces or dozens of grams of platinum. But with the steep price of platinum, this amounts to $3,000 or more, about as much as the cost of a whole gasoline engine. Searching for platinum alternatives is an active area of research, but so far most of the candidates are pricey as well.

Researchers are looking at ways to reduce the amount of platinum yet retain the catalytic activity. Peter Strasser, a researcher at the University of Houston in Texas, and his colleagues are trying to develop a platinum alloy that will do the job. An alloy such as bronze is a combination of elements, which in the case of bronze are tin and copper. Engineers often use alloys because they offer properties that are superior to those of a single metal, as described in chapter 1. A platinum alloy that acts as an effective catalyst in fuel cell electrodes yet contains less platinum would save a substantial amount of money.

As reported at Science*Daily* in 2007, Strasser and his colleagues discovered an alloy of copper (Cu), cobalt (Co) and platinum (Pt) that exceeded even pure platinum's catalytic activity for the reduction of oxygen (one of the important reactions taking place in fuel cells). The material consists of tiny particles called nanoparticles, described in chapter 2, which have a shell and an interior. This material is deposited on carbon electrodes. Strasser and his coworkers discovered that sending a varying electric current through the electrode separated most of the other metals from the shell, leaving the nanoparticle with mostly platinum on the surface. Surfaces are important because this is where a lot of catalytic activity takes place, and a collection of nanoparticles has a lot more surface area than the same amount of material in bulk. Yet Strasser believes the increase in surface area is not enough to account for all of the elevated activity rates, and the structure of the nanoparticle surfaces may contribute as well. Further research into this process may produce even greater efficiencies.

Another issue with platinum catalysts is that their capacity sometimes fades over time. Several factors are responsible, including a phenomenon similar to the side effects described for medications in chapter 3. Side effects occur when a medication acts on healthy tissue instead of the intended target. With platinum electrodes, the problem is that sometimes unwanted reactions occur at the electrodes. In the oxygen reactions taking place at the cathode, for example, hydroxide (OH) and other molecules sometimes form and bind to the platinum atoms. These molecules cover the platinum atoms and block access to the desired reactant, thereby reducing the catalytic activity. Sometimes the molecules even pull platinum atoms away from the surface, causing serious electrode degradation.

This problem is exacerbated when power requirements fluctuate. For instance, a fuel cell in an automobile would experience frequent stops, especially in city traffic, and platinum electrodes would rapidly lose their catalytic function.

But now researchers at Brookhaven National Laboratory in New York have found a method to stabilize platinum electrodes. The researchers used platinum nanoparticles as electrodes but modified the nanoparticles with the addition of thin layer of gold (Au). With scanning tunneling microscope (STM), described in chapter 2, and X-ray analysis, described in chapter 1, the research team determined that the gold formed clusters that protected the platinum from the attack of oxides. The gold-plated catalysts performed well more than 30,000 cycles of voltages varying from 0.6 to 1.1 volts. Junliang Zhang, Kotaro Sasaki, Eli Sutter, and Radoslav Adzic published this research in a paper, "Stabilization of Platinum Oxygen-Reduction Electrocatalysts Using Gold Clusters," in a 2007 issue of

Science. As the researchers noted, "There were insignificant changes in the activity and surface area of Au-modified Pt over the course of cycling, in contrast to sizable losses observed with the pure Pt catalyst under the same conditions."

CARS POWERED BY FUEL CELLS

With continued improvement, fuel cells having stable and relatively inexpensive electrodes will enable the development of practical fuel cell vehicles. If the fuel is hydrogen, obtained by some clean process, these will be zero-emission, environmentally friendly vehicles.

The importance of lowering automobile emissions has already resulted in changes in car manufacturing. Electric cars, which were popular 100 years ago but fell by the wayside because they could not keep up with gasoline-powered cars, are making a comeback. Although battery-powered vehicles are useful for short, slow trips around a crowded campus, for instance, they continue to be plagued by limited speeds and durations. Automobile manufacturers have alleviated these problems by com-

Experimental fuel cell powered bus *(NASA)*

bining battery power with a small gasoline engine. The resulting car, called a hybrid electric vehicle (HEV), gets more miles out of a gallon of gasoline than a conventional automobile, which helps the motorist's finances as well as the environment. But HEVs are not zero-emission vehicles.

Early versions of fuel cell vehicles have already been built and are being tested. For example, National Renewable Energy Laboratory (NREL), one of the U.S. government's leading centers for energy research, is evaluating fuel cell vehicles deployed at Hickam Air Force Base, in Honolulu, Hawaii. Scientists at NREL are involved in a large number of projects aimed at researching, designing, and testing a variety of technologies to improve the nation's energy efficiency. The sidebar on pages 154–155 provides more details about this important research and development laboratory.

Conservation and energy efficiency is critical at all levels, but the federal government is the biggest consumer of energy in the United States, and military operations account for the bulk of this usage—the Department of Defense accounts for nearly 80 percent of federal government energy consumption. Increased efficiency in federal government activities would translate into big savings. One young airman who worked on the flight line at Hickam Air Force Base in the 1980s once decided to conserve fuel by walking out to the airplanes instead of taking a motor vehicle. His supervisors were favorably impressed with his initiative. This conservation effort worked out well, until one day when even the young airman's spry legs almost failed to get him out of the way of a taxiing KC-35 airplane, the pilot of which was for some reason in a hurry to reach the hangar. From then on, it was back to the motor vehicles. (But the airman, who now has a more sedate job—and is the author of this book—continues to walk instead of drive when he can safely do so.)

Two types of vehicle are being tested at Hickam Air Force Base. One is a 30-foot (9.1-m) bus, used for shuttling passengers around the base, and the other is a step van to carry packages and equipment (similar to the vans used by FedEx and UPS). Powering these vehicles are hybrid engines, consisting of both a battery pack and a fuel cell. The bus has a relatively small fuel cell stack, capable of generating 20 kilowatts, and relies more on batteries supplemented by fuel cell energy. But the step van contains a larger, 65-kilowatt fuel cell system. The fuel cells of both vehicles are proton exchange membrane cells operating on hydrogen.

To provide fuel, officials built a hydrogen station on the base. This station consists of a power control unit, a water container, a fuel processor that produces hydrogen by water electrolysis, a compressor (to

National Renewable Energy Laboratory

In 1977, as oil embargoes and skyrocketing gasoline prices called attention to America's dependence on foreign oil, the Solar Energy Research Institute began operations. The goal of this research center was to explore technologies that could make use of freely available energy from sunlight as

The National Renewable Energy Laboratory *[National Renewable Energy Laboratory [NREL]]*

compress the hydrogen for storage), and a storage unit. Hydrogen is a highly flammable gas—the German airship *Hindenburg* crashed in New Jersey on May 6, 1937, killing 36 people, when a spark ignited the ship's hydrogen. The Hickam hydrogen station has seven emergency shutdown devices, any of one which can shut down the station by removing power and closing valves.

an alternative to oil and gasoline. The research focus broadened in September 1991, when the institute became a national laboratory of the Department of Energy. At this time, the name changed to National Renewable Energy Laboratory (NREL). The main laboratory and offices are located at Golden, Colorado, a short distance from Denver.

With a staff of about 1,000, including researchers and engineers from diverse backgrounds in physics, chemistry, and biology, NREL investigates many possible sources of renewable energy. Renewable energy refers to sources of energy that are continually replenished, unlike sources such as oil or coal that are of limited and dwindling quantity. Programs include research on biomass (the use of agricultural or other biological material as fuel), geothermal sources (energy derived from Earth's heat), wind and water power, solar energy, fuel cells, and other technologies designed to increase the efficiency of buildings, energy distribution, industry, and transportation.

An important component of NREL strategy is its Technology Transfer Office. Even the most inventive ideas do little good if they are not developed beyond the laboratory stage. NREL works with companies and government agencies to help bring new technologies into the marketplace. For example, the fuel cell vehicles that NREL evaluated at Hickam Air Force Base were the products of a collaborative team from Enova Systems, Hydrogenics Corporation, and the Hawaii Center for Advanced Transportation Technologies.

The two fuel vehicles began operation late in the summer of 2007. Users maintain logs that document vehicle performance, and tests are ongoing to evaluate the operational experience and the refueling station. Other NREL evaluation projects involve transit buses powered by fuel cell hybrids in Oakland, California, Thousand Palms, California, and Hartford, Connecticut.

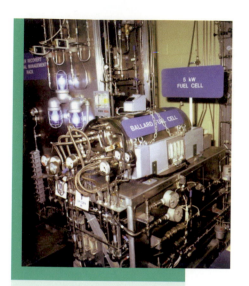

Experimental fuel cell for a bus *(NASA)*

Car manufacturers are also beginning to develop fuel cell technology for their products. Honda, a Japanese company, has been experimenting with fuel cells for several years and in 2002 began leasing a fuel cell vehicle. These early cars were limited and expensive models and were not intended to compete with gasoline- or hybrid-powered vehicles. But on November 14, 2007, Honda presented a new version of their FCX Clarity, a zero-emission fuel cell vehicle, the only by-product of which is water. The fuel cell has an electrolytic membrane and generates 100 kilowatts of power, and the vehicle has a range of 270 miles (430 km) with its hydrogen storage capacity. Honda leases the vehicle in the United States on a limited basis.

Zero-emission vehicles such as the buses and vans tested by NREL and Honda's FCX highlight the progress that people are making in applying fuel cell research to solve real problems. But despite continued innovation, difficulties remain. These vehicles are expensive and beyond the budget of many consumers. There is also a problem of obtaining fuel. Finding a gasoline station and filling up a car's tank is no trouble except for desolate stretches of road such as in Death Valley, California. In contrast, few stations exist to service a hydrogen-powered vehicle. Solving the supply problem with hydrogen fuel cells is essential before the United States and other countries can enjoy the benefits of having zero-emission cars on the road.

HYDROGEN ECONOMY

The term *hydrogen economy* refers to a situation in which most of the population's energy needs are met with hydrogen fuel instead of oil and coal. In this scenario, the economy will rely on hydrogen to power vehicles, produce electricity, and other essential activities.

Iceland is already taking the first steps in this process. This island nation of about 300,000 people has a penchant for self-sufficiency in terms of energy, deriving the bulk of their energy from hydropower (energy from the flow or fall of water) and geothermal energy, which uses Earth's heat. Iceland began operating a few hydrogen buses in Reykjavík, the capital and largest city, in 2003. In order to fuel these buses, and also to prepare for a future in which hydrogen-powered vehicles become more plentiful, the first public hydrogen refueling station opened on April 24, 2003.

As hydrogen fuel cells become more common, the demand for hydrogen will grow. But fuel cells are not the only method of obtaining energy from hydrogen. Hydrogen can be burned in combustion engines, similar to gasoline. All heat engines suffer from the constraints of thermodynamics, but burning hydrogen instead of gasoline produces water instead of harmful emissions. The German manufacturer BMW has been working with engines that can run on hydrogen since 1979, when the company built an engine capable of burning either gasoline or hydrogen. In 2004, BMW made the H2R, a sleek hydrogen-powered racer that has an engine capable of generating 232 horsepower—173 kilowatts—and the car can reach a speed of 187 miles per hour (300 km/hr).

But the zero-emission promise of a hydrogen economy, whether it comes from highly efficient fuel cells or less efficient hydrogen combustion, cannot be fulfilled solely by the development of hydrogen-powered vehicles. The problem is that there are no major sources of hydrogen readily available. This is not to say that there is little hydrogen—it is by far the most abundant element in the universe and is found in a lot of compounds on Earth, such as water. Methods such as electrolysis to extract hydrogen from these compounds are not cheap and require energy. If, for example, the electricity needed for electrolysis comes from coal- or oil-powered generators, which emit pollution, the zero-emission benefit of hydrogen fuel cells is compromised. Although fuel cell operation emits only water, air pollution was generated in the production of the hydrogen fuel.

Iceland escapes this dilemma because most of their energy is from "clean" sources, hydropower and geothermal, rather than from hydrocarbon combustion. Since there are few emissions in the production of their electricity, providing hydrogen by electrolysis does not spoil the zero-emission advantage for Iceland's fuel cells.

But using electricity to generate hydrogen, as in electrolysis, is quite expensive. Hydrogen is more commonly made in the rest of the world by a process known as steam reforming. This process involves heating a hydrogen-containing substance such as methanol (CH_3OH) or natural gas (which is mostly methane, CH_4) with water in the presence of a catalyst. The reaction releases hydrogen gas (H_2) and carbon monoxide, most of which is then converted to carbon dioxide. Impurities in the mixture result in pollutants in the product mix as well as carbon dioxide, a greenhouse gas.

Researchers are addressing this issue by improving the process's efficiency. Hydrogen must be filtered out of the reaction products, which requires substantial energy and often the use of poisonous chemicals. A more effective, lower-energy solution would employ a polymer (plastic) membrane developed by Haiqing Lin, Elizabeth Van Wagner, and Benny Freeman at the University of Texas in Austin and their colleagues from the Center for Energy Technology in Research Triangle Park, North Carolina. The membrane works by trapping hydrogen molecules; the other reactant products such as carbon dioxide interact with the polymer and pass through the thin film. The researchers published their paper, "Plasticization-Enhanced Hydrogen Purification Using Polymeric Membranes," in a 2006 issue of *Science.*

Even small increases in efficiency can help. If fuel cell vehicles ever become the predominate means of transportation, hydrogen demand will soar. Advances in hydrogen production and purification will ease the burden and help make these processes as environmentally friendly as possible.

CONCLUSION

Motivated by the need to reduce pollution and greenhouse gases, along with the desire to develop sustainable, renewable sources of energy, fuel cell research has blossomed in recent years. This research has taken several paths, including the search for less expensive catalysts and the perfection of techniques to produce and purify hydrogen from readily available compounds. If the aim of zero-emission transportation is to be achieved, no step in the energy conversion process—from the production of fuel to its use in the fuel cell—can be allowed to contribute to pollution or climate change.

As the costs of fuel cells go down, more consumers will be able to afford to buy them. If advances in hydrogen production can keep pace, people can enjoy the benefits of motorized technology without putting undue pressure on the environment.

One optimistic scenario would tie hydrogen production to the ripening development of alternative sources of energy. Iceland, for example, makes use of their volcanic geology that provides a considerable amount of extractable heat energy—the geothermal source mentioned earlier. In the United States, NREL is pursuing a number of options for alternatives to depletable and polluting hydrocarbon combustion sources, as described in the sidebar on pages 154–155. Harnessing energy from sunlight, tides, wind, waterfalls, and Earth's hot interior are all options to be explored.

A clean and renewable energy source would benefit the economy and the environment in many ways but would not solve all of the problems. The energy must be used efficiently, and this is where technology such as fuel cells enters the picture. A clean and renewable energy source, if developed, could produce enough hydrogen to satisfy the growing demand as fuel cells become more affordable and popular. Fuel cells would be a convenient way to use the energy output—adding another environmentally friendly step in the energy conversion chain. In such a hydrogen economy, hydrogen would be the medium, or carrier, of energy from the natural source to the motors or vehicles that make the economy go.

With little danger of running out of hydrogen—it is not destroyed in the process, as the product of fuel cells is H_2O—energy from this source can be used in a huge number of ways if fuel cell research is successful. For example, Boeing researchers began working on the Fuel Cell Demonstrator Airplane in 2003. The airplane will use a hybrid power system—a proton exchange membrane fuel cell along with a lithium ion battery—which will drive a regular propeller. As the airplane cruises in midair, the fuel cell will supply the power, but for climbing and takeoff the battery pack will supplement the output.

Hydrogen-powered flight will probably not be common in the near future, but Boeing is showing that it is feasible. As the fuel cell research frontier expands, so do the potential applications.

CHRONOLOGY

1766 C.E.	British chemist Henry Cavendish (1731–1810) discovers hydrogen, which he calls "inflammable air."

1800   Italian scientist Alessandro Volta (1745–1827) invents the battery.

British chemists William Nicholson (1753–1815) and Sir Anthony Carlisle (1768–1842) discover electrolysis of water.

1834   British scientist Michael Faraday (1791–1867) describes the scientific principles of electrolysis.

1838   Welsh lawyer Sir William Robert Grove (1811–96) invents a hydrogen fuel cell.

1839   Swiss chemist Christian Schönbein (1799–1868) publishes the first description of a fuel cell.

1889   German chemist Ludwig Mond (1839–1909) and his colleagues experiment with hydrogen cells and use the modern term *fuel cell*.

1959   British engineer Francis Thomas Bacon demonstrates the first practical fuel cell.

1960s   The Apollo missions, the United States manned space missions that explored the Moon, use fuel cells to provide electricity to the spaceship as well as water for the astronauts.

General Electric scientist William Grubb and his colleagues develop a proton exchange membrane fuel cell.

1966   General Motors, an American manufacturer, produces a vehicle called the Electrovan that is propelled by a hydrogen fuel cell, generally considered the first fuel cell vehicle. Far too expensive for consumer use, the vehicle was not marketed and the project was discontinued.

1981   NASA launches the first space shuttle, which is equipped with fuel cells.

1990s	Ballard Power Systems, Inc., a Canadian company, develops efficient and practical proton exchange membrane fuel cells.
2002	Honda, a Japanese manufacturer, develops a practical fuel cell vehicle.
2003	Iceland opens a public hydrogen refueling station.
2005	The Energy Policy Act of 2005 encourages the use and advancement of clean sources of energy by providing tax breaks and other incentives.
2007	Peter Strasser and his colleagues discover an alloy of copper, cobalt, and platinum that exceeds pure platinum's catalytic activity for the reduction of oxygen (one of the important reactions taking place in fuel cells).
	National Renewable Energy Laboratory begins evaluating fuel cell vehicles deployed at Hickam Air Force Base, in Honolulu, Hawaii.

FURTHER RESOURCES

Print and Internet

Ewing, Rex A. *Hydrogen—Hot Stuff Cool Science,* 2nd ed. Masonville, Colo.: PixyJack Press, 2007. This lighthearted look at hydrogen includes chapters on the chemistry of hydrogen as well as electrolysis, hydrogen storage, and fuel cells.

Gibilisco, Stan. *Alternative Energy Demystified.* New York: McGraw-Hill, 2007. Gibilisco briefly explains the concept of energy and describes the variety of ways in which it is generated. Included is a chapter on electric, hybrid, and fuel cell vehicles.

Hoffmann, Peter. *Tomorrow's Energy: Hydrogen, Fuel Cells, and the Prospects for a Cleaner Planet.* Cambridge, Mass.: MIT Press, 2002. This book describes the history of hydrogen-powered projects and discusses the safety, economics, and research of today, including fuel cell vehicles.

Intergovernmental Panel on Climate Change. *Climate Change 2007: Synthesis Report.* Available online. URL: http://www.ipcc.ch/pdf/assessment-report/ar4/syr/ar4_syr.pdf. Accessed May 28, 2009. IPCC scientists review climate data and models.

Lin, Haiqing, Elizabeth Van Wagner, Benny D. Freeman, Lora G. Toy, and Raghubir P. Gupta. "Plasticization-Enhanced Hydrogen Purification Using Polymeric Membranes." *Science* 311 (February 3, 2006): 639–642. The researchers developed a membrane that traps hydrogen molecules, letting other reactant products, such as carbon dioxide, pass through.

Lower, Stephen. "All About Electrochemistry." Available online. URL: http://www.chem1.com/acad/webtext/elchem/. Accessed May 28, 2009. Part of a virtual chemistry textbook, this excellent resource explains the basics of electrochemistry, which is important in understanding how fuel cells work. Discussions include galvanic cells and electrodes, cell potentials and thermodynamics, the Nernst equation and its applications, batteries and fuel cells, electrochemical corrosion, and electrolytic cells and electrolysis.

Romm, Joseph J. *The Hype About Hydrogen: Fact and Fiction in the Race to Save the Climate.* Washington, D.C.: Island Press, 2004. All new technologies have their doubters and pessimists. Romm, formerly employed at the Department of Energy during President Clinton's administration, is not quite convinced that convenient and practical means of getting energy from hydrogen will soon be forthcoming.

Science *Daily.* "New Class of Catalyst for Fuel Cells Beats Pure Platinum by a Mile." News release, October 24, 2007. Available online. URL: http://www.sciencedaily.com/releases/2007/10/071023164031.htm. Accessed May 28, 2009. The researchers have discovered an alloy of copper, cobalt, and platinum that exceeds pure platinum's catalytic activity for the reduction of oxygen.

Smithsonian Institution. "Fuel Cell Basics." Available online. URL: http://americanhistory.si.edu/fuelcells/basics.htm. Accessed May 28, 2009. This Web page presents an overview of fuel cell operation. Alkali, molten carbonate, phosphoric, proton exchange membrane, and solid oxide fuel cells are discussed.

Zhang, J., K. Sasaki, E. Sutter, and R. R. Adzic. "Stabilization of Platinum Oxygen-Reduction Electrocatalysts Using Gold Clusters." *Sci-*

ence 315 (January 12, 2007): 220–222. The researchers found that certain gold clusters protect platinum from the attack of oxides.

Web Sites

Breakthrough Technologies Institute: Fuel Cells 2000. Available online. URL: http://www.fuelcells.org/. Accessed May 28, 2009. Breakthrough Technologies Institute, a nonprofit organization devoted to promoting fuel cell research and development, sponsors this informative Web site. Sections include fuel cell basics, hydrogen, and the latest research and industrial news.

Department of Energy: Hydrogen, Fuel Cells & Infrastructure Technologies Program—Education. Available online. URL: http://www1.eere.energy.gov/hydrogenandfuelcells/education/h2iq.html. Accessed May 28, 2009. This Web site is loaded with links to fact sheets, documents, and animations that explain in accessible terms the technology and operation of fuel cells.

National Fuel Cell Research Center. Available online. URL: http://www.nfcrc.uci.edu/. Accessed May 28, 2009. Based at the University of California, Irvine, the National Fuel Cell Research Center investigates fuel cell systems and the strategies that will allow businesses to take advantage of this technology. The center also develops educational resources such as tutorials, answers to frequently asked questions (FAQs), and other important information included on their Web site.

National Renewable Energy Laboratory: Hydrogen and Fuel Cells Research. Available online. URL: http://www.nrel.gov/hydrogen/. Accessed May 28, 2009. National Renewable Energy Laboratory is a national laboratory in the United States devoted to exploring sustainable sources of energy. This Web site describes the laboratory's investment in hydrogen and fuel cells. Projects include hydrogen production and delivery, storage, fuel cells, and safety.

NOVA: Fuel Cells. Available online. URL: http://www.pbs.org/wgbh/nova/sciencenow/3210/01.html. Accessed May 28, 2009. PBS airs *NOVA*, a series that explores a variety of science and technology topics. The companion Web site also provides a great deal of information, as in this Web page, which offers a debate about the future of hydrogen fuel cells, along with a view under the hood of a fuel cell car.

6

Archaeological Chemistry—Exploring History with Chemistry

The French philosopher Auguste Comte (1798–1857) was wary of speculations—he did not like ideas with little support from scientific experimentation or observation. In his multivolume work *Cours de Philosophie Positive* (*Course of Positive Philosophy*), published in 1830–42, Comte cited an example of a question that he felt would be forever speculative. Because the planets and stars are so distant, Comte believed that people would never gain firm knowledge of the composition of these astronomical objects. Yet only a few years later, scientists learned about spectroscopy and how to determine the elements that compose stars and planets by analyzing emitted or reflected light. Comte's prediction was soon proven untrue.

A similar prediction could easily be made in regard to the study of history. Events in recent times are well documented, but earlier times are much sketchier, with only a few *artifacts* remaining. Short of constructing a time machine, which is not possible (at least not yet), a detailed knowledge of the distant past would seem unattainable.

But thanks to a frontier of chemistry described in this chapter, people are gaining an increasing amount of knowledge about the distant past. These achievements are satisfying because they advance knowledge and resolve long-standing mysteries. But this knowledge may also have a great deal of practical value. As the philosopher George Santayana (1863–1952) noted in 1905, "Those who cannot remember the past are condemned to repeat it." Ancient civilizations such as Rome rose to greatness and later collapsed, which makes historians ponder how and why these people were so successful, yet ran into problems they either did not recognize or failed to solve.

Modern society is also facing obstacles such as pollution, scarcity of resources, global climate change, and others. Some of these problems are difficult to understand and may require a response that is not easy to implement. An understanding of past errors may give people of to-day vital clues to avoid similar collapses and catastrophes. This chapter examines the methods that chemists and archaeologists are developing to learn how people of long ago lived and died.

INTRODUCTION

Archaeology is the study of the remains of past human life and culture. The term *archaeology* comes from Greek words meaning ancient knowledge, and archaeologists often study people and societies that thrived long ago and left little or no written record—such people left no history written by their own hand. Yet even in cases where some written records survive, archaeologists also learn much by studying the tools, structures, utensils, weapons, and biological materials of these people.

One of the chief tools of an archaeologist is the shovel, or some other means of digging. Soil covers most artifacts over time, so objects of the past, even if they were used and left on the surface, eventually get buried. Finding these artifacts can be difficult. In some cases, such as mounds of earth built by Native Americans, there are changes on the surface that indicate an important archaeological site. (A number of Native American cultures constructed mounds, sometimes for burial but also for a variety of other purposes.) But often all visible traces are destroyed by weather, or covered by dirt and the activity of subsequent inhabitants.

Many archaeological sites are discovered by accident. Rome, for example, is a city in Italy that has been occupied for nearly 3,000 years. This city was the cultural and political center of the Romans, a people whose

Archaeologists working at the Santul Mare site in Romania
(Rechitansorin/Dreamstime.com)

empire stretched over much of the area surrounding the Mediterranean Sea 2,000 years ago. Some of the structures from Roman times survive today, such as the Colosseum, which is a huge arena that once seated 50,000 spectators, and aqueducts that carried water to Rome from the surrounding area. Other remains of this long-ago era are buried under the houses, theaters, office buildings, and roads of modern Rome. (Much of the city today is about 20–50 feet [6–15 m] higher than in ancient times because it is built on the rubble of the past.) Workers digging at a new construction site in Rome often encounter an archaeological surprise. For example, during work to create an office building in 2006, workers found a temple of the second century C.E. dedicated to Emperor Hadrian's mother-in-law.

Rome has a fascinating history, which is known to a certain extent through surviving documents written in Latin. Knowing the language helps tremendously in interpreting archaeological finds. In other cases, where records are sparse, confusion often develops. Many layers pile on top of one another at sites having lengthy periods of occupation, as in Rome, and archaeologists may have to sort through the layers without having a map or a listing of who lived where and at what time.

Ruins of the ancient city known as Troy, for example, are believed to lie beneath the rubble of later cities at a site called Hissarlik in Turkey. These multiple layers confused the German archaeologist Heinrich Schliemann (1822–90) as he excavated the site in the 19th century, searching for relics of the famous Trojan War with the Greeks. This war is described in the epic *Iliad,* attributed to the Greek poet Homer, but its historical accuracy is debatable. Further excavations of this site have found evidence of warfare around the time the events in the *Iliad* supposedly took place, probably 3,000–3,500 years ago. But this is hardly surprising, since war was common in this era. No one knows if the Trojan War as related by Homer really happened, and the ancient Trojans cannot offer much help—their language is currently unknown.

Egypt is the site of another civilization that has a long and glorious past. The pyramids at Giza stand as prominent reminders of a venerable culture, and they are so old that ancient Greeks such as Herodotus (ca. 484–425 B.C.E.) marveled at them. But the language carved on monuments or found on old documents made of papyrus was a mysterious set of symbols known as hieroglyphics. Researchers made great strides in understanding ancient Egypt when a French Army officer discovered a stone in 1799 at Rosetta (Rashid), an Egyptian port city. This stone contained a passage in three different languages—hieroglyphics, demotic script (another type of Egyptian writing), and Greek. The passage described an edict issued in 196 B.C.E. Since linguists can understand Greek, the stone provided a key to translating hieroglyphics.

Written records provide a valuable glimpse into the thoughts and behavior of ancient peoples. But these records are not complete and omit important information (important, that is, to historians). Scientists studying the past have had to turn to other sources such as analyzing artifacts. Archaeological finds are a rich source of information, if that information can be unlocked. The need is to find a "Rosetta stone" for other types of artifacts—a method of gleaning all the information that is available.

Chemistry offers a reliable method of extracting information from a large variety of archaeological finds. Some of the most important of these are human remains. Human remains from ancient times are usually skeletons, as the soft tissues of the body decompose quickly. But in Egypt and a few other places, archaeologists have discovered more complete samples. Wealthy Egyptians, believing in an afterlife that required maintenance of the body, were often preserved after death. These mummies exhibit the remarkable technological skills of the ancient Egyptians, as some of the remains have survived for several thousand years. Other intentionally preserved human remains have been found in China.

The very existence of these mummies provides clues as to the religion, philosophy, and technology of these ancient civilizations. Along with the surviving records, such as the Book of the Dead, which described the passage of the deceased into the afterlife, historians know something about ancient Egyptians and their beliefs.

But what about daily life? Religious or political writings do not usually reveal much about how many of the citizens actually lived and died. To investigate this issue, archaeologists employ the principles of chemistry and biology to study the bodies. These investigations are similar to forensic anthropology—the application of anthropology (the study of humans) in legal or criminal situations.

A COLD CASE: ÖTZI, THE ICEMAN

Intentional preservation such as that of Egyptian and Chinese mummies is not the only means by which archaeologists find ancient bodies. Sometimes accidental circumstances set up environmental conditions that also preserve a body. For example, peat bogs are wetlands with an accumulation of peat—partially decayed plants—and are common in northern Europe. Bodies that fell or were thrown into peat bogs thousands of years ago have been preserved, possibly due to the bog's lack of oxygen or the presence of antimicrobial chemicals. Hundreds of "bog mummies," and partial mummies, have been found.

Ice is another preservative. Freezing preserves bodies by slowing or stopping chemical reactions that break down tissues. One of the best preserved bodies from ancient times was found on September 19, 1991, when German tourists hiking at 10,530 feet (3,210 m) above sea level in an isolated section of the Ötzal Alps, on the border between Italy and Austria, discovered a body that came to be known as Iceman, or Ötzi (after the mountain range). (Since Tyrol is another name for the region in which Ötzi was found, some people refer to him as the Tyrolean iceman.) The body was clothed in a coat made from woven grass, a leather vest, leggings, and shoes. His shoes were particularly impressive—they were waterproof and consisted of bearskin, deerskin,

Ötzi the Iceman, and some artifacts found around his body
(*Werner Nosko/EPA/Corbis*)

and tree bark. Among his effects were a copper axe, a flint knife, and a quiver of arrows, although most of the arrows were not finished. The body was shrunken and hairless—the hair and outer skin had peeled off the body during its long stay in the ice.

When did this man die? Although the man had clearly died some time ago, police still could have opened an investigation of a crime committed at some time in the past, called a cold case since the clues and the trail of the perpetrator would not be fresh. (Considering the condition of the body, the term applies in more ways than one.) But radiocarbon dating established a time of death that rendered any judicial action a futile exercise: Ötzi did not die a month or a year ago, or any time in the modern era—he died about 5,300 years in the past, in a period known as Neolithic (New Stone Age). This makes Ötzi the oldest intact mummy ever found as of May 2009. Ötzi's age is startling, but radiocarbon dating, as described in the following sidebar, is a reliable means of dating organic material.

Radiocarbon dating reveals when Ötzi died, but not how old he was. Forensic anthropologists can approximate the age at death by studying

Radiocarbon Dating

All atoms of the same element have the same number of protons in the nucleus—the atomic number—but some atoms have a different number of neutrons, giving rise to different isotopes. Although isotopes of a given element have chemical properties that are very nearly identical, the number of neutrons in the nucleus can cause instability, leading to the emission of particles and a decay, known as radioactive decay, of the nucleus. In this process, the composition of the nucleus changes, resulting in a different isotope or an isotope of a different element. For example, carbon 14 (which has 14 particles in the nucleus—six protons and eight neutrons) decays into nitrogen 14 (seven protons and seven neutrons) when a neutron in carbon 14 converts into a proton and an electron (which is emitted).

Carbon is a critical element in biology since the element plays a vital role in the formation of long molecular chains such as DNA and protein, which are essential for all living organisms on the planet. Almost all of the carbon compounds on Earth and in the atmosphere consists of stable carbon isotopes, mostly carbon 12 and a little carbon 13, which do not decay. But high-energy protons called cosmic rays constantly bombard the planet; although astronomers are not sure of the source of these speedy particles, they exert considerable effects, including collisions with nitrogen 14 atoms in the upper atmosphere that produce unstable carbon 14. (Most of the reactions occur at high altitudes since only a few cosmic rays reach the ground.) The decay of carbon 14 occurs spontaneously, at random, and is unaffected by environmental conditions. But even though the decay is random, the probability—the likelihood that decay will occur—is constant, and a group of atoms decays at a specific and measurable rate. Half of a given sample of carbon 14 atoms will decay in 5,730 years—this period is called the *half-life*—and half of this remainder will decay in another 5,730

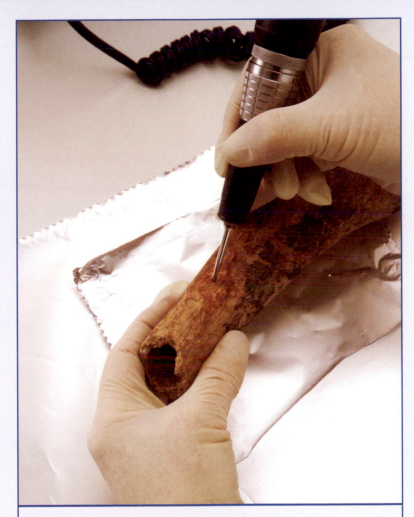

Researcher removing a sample from a bone for radiocarbon dating
(James King-Holmes/Photo Researchers, Inc.)

years. So after 5,730 + 5,730 = 11,460 years, 25 percent of the original carbon 14 atoms remain, while 75 percent has decayed into nitrogen 14.

(continues)

(continued)

About one atom in a few trillion carbon atoms is carbon 14, but these isotopes drift around and are taken up by living organisms, including humans as they eat, and by plants during photosynthesis. Some of the carbon 14 atoms decay but are replaced during the organism's food intake. When the organisms die, the carbon 14 in the body decays and is not replaced. Measurements of the carbon 14 to carbon 12 ratio indicates how long the body stopped taking up carbon 14—or, in other words, how long ago the organism died. The University of Chicago researcher Willard Libby developed this procedure, called radiocarbon dating, in 1949.

Archaeologists widely use radiocarbon dating as a means of dating organic material, but the technique has some caveats. In order to determine how much the carbon 14/carbon 12 ratio has declined, archaeologists must know what this ratio was at the time of the organism's death. The value of this ratio should be roughly the same as the ratio in the atmosphere, but the atmospheric ratio of carbon 14/carbon 12 has changed slightly over time, so archaeologists must take this change into account. The sample must not be contaminated with any new carbon, which could introduce more carbon 14 and throw off the ratio. Radiocarbon dating is also useful only for the recent past; beyond about 50,000 years, so much carbon 14 has decayed that the tiny residue is almost impossible to measure accurately. Ötzi is well within this range, for he died, in terms of carbon 14, only a little less than one half-life ago.

wearing and aging of the teeth and bones. Although it is difficult to be precise, Ötzi was about 40–45 years old at the time of his death. Considering the ravages of disease and the precarious existence of human beings in this era, Ötzi was probably considerably older than many of his contemporaries.

The mummy only weighs about 30 pounds (13 kg) and is about five feet (1.54 m) in length, but because of shrinkage this does not reflect the Iceman's true physique. According to the South Tyrol Museum of Archaeology in Bolzano, Italy, which houses the body, the living Ötzi would have been about 63 inches tall (1.6 m) and weighed 110 pounds (50 kg).

But where did he come from? He may have been raised in the same region where his body was found, or he might have been a newcomer to the area, or just traveling through. To determine which is the case, chemists and biologists have developed techniques that make use of certain isotopes. As with forensic evidence, these isotopes can be found in Ötzi's teeth and bones.

A tooth consists of dentin, which gives the tooth its shape and size, and a covering of enamel, which is extremely hard—enamel is the hardest material in the body—and protects the tooth from the grinding and chewing associated with eating. (Even so, some of it wears away over the years, providing an indication of age.) Both dentin and enamel are primarily made of calcium phosphate, but there are differences in how the body generates these two structures. Enamel forms only once, when the tooth initially grows, and has to last a lifetime. Dentin, however, gets slowly replaced throughout life. Bone is similar to dentin and also experiences gradual turnover.

Because tooth enamel forms only once, its structure supplies a "snapshot" of the time of tooth formation, which occurs early in life. The tissues and structures of the body are built from elements supplied by food and water. For enamel, these elements consist of those that were available during childhood. The key to identifying where the Iceman spent his childhood comes from the variation in isotopes that occur in different regions.

Calcium is an important element in teeth and bone, but strontium is chemically similar and can replace calcium in these structures. This is important to archaeologists because certain strontium isotope ratios vary in soils—and therefore vary in the foodstuffs grown in those soils. For example, the ratio of stable isotopes strontium 87 to strontium 86 can be a distinct marker for certain soils. Another important element is oxygen, which is a major component of calcium phosphate. Oxygen can be incorporated into the body from food or water, and ratios of oxygen 18 to oxygen 16 often distinguish geographic location.

By examining Ötzi's teeth and bones and comparing their isotope ratios to those found in the nearby environments, archaeologists can get clues about where the Iceman grew up and where he spent his adulthood. (More precisely, these measurements indicate where Ötzi's food and water came from. But in this ancient era, Ötzi would have probably obtained most or perhaps all of the necessities of life from the local area.) Teeth enamel provide clues of childhood environment, and bone, which gets remade every decade or two, indicates a more recent abode.

This map shows where the Iceman was found and the valleys in which he lived.

Chemists use an instrument called a mass spectrometer to measure the relative abundance of isotopes. There are different kinds of mass spectrometers, but the basic idea is to measure the mass of a substance by applying a force. The response to this force depends on the object's mass—think of Newton's second law, where acceleration equals force divided by mass. In the case of mass spectroscopy, the substances to be measured are first ionized—they are made into charged particles called ions by stripping electrons. A magnetic field deflects the motion of an ion, and the deflection depends on the ion's mass, most of which is due to the protons and neutrons in the nucleus. The technique separates different isotopes and measures their abundance in a given sample.

In 2003, Wolfgang Müller, a researcher then at the Australian National University in Canberra, and his colleagues performed isotope measurements of Ötzi's teeth and bones, then compared these values with surrounding soils, rocks, and streams. In order for these technique to succeed in pinpointing locations, there must be some variability—a uniform environment would give no clues—and fortunately there is variation in the complex alpine forests and valleys of the Ötzal Alps. Müller and his colleagues found that the Iceman probably spent his childhood in the Eisack Valley, located 25–30 miles (40–48 km) from the site he was found, as shown on the map opposite. As an adult, he spent considerable time in the Etsch Valley, about 12 miles (19.2 km) south of where he died. Based on this evidence, Ötzi did not stray too far from home. Müller and his colleagues published their findings in a paper, "Origin and Migration of the Alpine Iceman," in a 2003 issue of *Science.*

Other clues shed light on Ötzi's culture. As Klaus Oeggl, a researcher at Innsbruck University in Austria, and his colleagues noted in a 2007 issue of *Quaternary Science Reviews,* "The investigations of the Tyrolean Iceman 'Ötzi' and his artefacts, discovered at a remote location high in the Eastern Alps, have contributed greatly to the knowledge of the lifestyle of Neolithic humankind." The Iceman was well clothed to survive in the harsh alpine environment, and the blade of his copper axe was extremely pure and well made. He was not in the best of health, for he suffered from arthritis, as evidenced by the condition of his joints, and the remains of parasitic whipworm eggs found in his intestine indicate he would have been experiencing digestion trouble.

Further information about a person's background is contained in the hereditary material known as deoxyribonucleic acid (DNA). DNA consists of long sequences of covalently bonded molecules known as nucleotides, or bases—adenine (A), thymine (T), cytosine (C), and guanine (G)—that provide the blueprint for the body and all of its metabolic activities. Genes are inherited from parents, but because of the large number of genes in humans—about 25,000—and the mixing that occurs during reproduction, each person has a unique sequence. (Exceptions are identical twins, whose DNA sequences are the same because identical twins come from the same fertilized egg cell.) Forensics experts often match DNA sequences found in blood or tissue samples from a crime scene to that of a suspect, placing him or her at the scene of the crime. In principle, DNA from the Iceman and other ancient sources could reveal a great deal of information. The problem with the use of DNA in archaeological chemistry is that it rapidly deteriorates.

ANCIENT DNA

A DNA molecule is normally a double-stranded helix, with the strands held together by relatively weak hydrogen bonds. The structure and geometry of the nucleotide bases are such that A on one strand only pairs with T on the opposite strand, and C pairs with G. This arrangement is known as complementary binding—A and T are complementary, as are C and G. A DNA helix is stable, but the strands are not joined so tightly that they cannot be "unzipped," which must occur when genes are read or the DNA is copied during cell division.

A cell's DNA is contained and protected by little compartments that are surrounded by membranes. Enzymes known as nucleases chop up any free DNA floating around the cell; these enzymes act as cellular sentinels, guarding against invading viruses that try to sneak their genes into the cell. When an organism dies, the cellular membranes break down and nucleases destroy much of the cell's own DNA. Finding DNA in a body that has been around for a while is unlikely.

But certain conditions such as freezing or desiccation (removal of water) can greatly reduce the activity of these enzymes. This means that mummies such as Iceman, despite its great age, may contain some quantity of DNA in good enough shape to determine its sequence. There are other enemies of ancient DNA, though, including radiation, which breaks apart the strands, and chemical reactions such as oxidation and hydrolysis. Over time, radiation and chemical reactions damage the DNA so that its sequence is undecipherable.

Two compartments in the cells of humans (and other animals) have DNA—the nucleus, which contains most of the genes of the organism, and mitochondria. Mitochondria are membrane-bound compartments scattered throughout the cell and are involved in extracting energy from food molecules by a series of complex chemical reactions. Capable of dividing, mitochondria resemble bacteria and have their own DNA. (Most biologists believe that mitochondria were once independent organisms that were captured and assimilated into ancestors of animal cells. The Russian scientist Konstantin Mereschkowsky [1855–1921] first proposed this theory in the early 20th century.) Genes in mitochondrial DNA (mtDNA) produce some but not all of the proteins needed by mitochondria—the rest come from genes in nuclear DNA.

While each cell has only one nucleus, some cells require a lot of energy and have up to several thousand mitochondria. Because of these multiple copies, bits and pieces of mtDNA are more likely to survive than nuclear DNA.

A few years after Ötzi was found, Svante Pääbo, then at the University of Munich in Germany, and his colleagues searched for DNA in eight small tissue samples of muscle, connective tissue, and bone, coming from the left hip area. Because the amount of DNA was so tiny, the researchers had to amplify the DNA molecules by making many copies of them, otherwise they would never have been able to find and sequence them. DNA amplification can be done in a technique known as polymerase chain reaction (PCR). This technique has become a standard procedure for forensics experts sifting through crime scene samples, researchers who sequence genomes (complete sets of genes for an organism), and archaeological chemists hunting for ancient DNA. The following sidebar describes PCR in more detail.

Although the Iceman lived a long time ago, his physique and his biology seem identical to modern humans, so Ötzi's genes should be quite similar to modern humans. This helped Pääbo and his colleagues to choose the appropriate sequence for the primers, which as described in the sidebar on PCR are needed to initiate the DNA amplification process. The choice of primer is essential because this is the beginning point of the copying, and it defines what DNA segments will be amplified. A primer has to bind in order to provide the starting point, and only complementary sequences will bind tightly enough. Pääbo doubted that any

Polymerase Chain Reaction

Kary Mullis, the inventor of polymerase chain reaction (PCR), earned a Ph.D. in biochemistry from the University of California, Berkeley, in 1972. The late 1960s and early 1970s were turbulent times in Berkeley, with Vietnam war protests, civil rights activism, and certain people—"hippies"—who engaged in controversial lifestyles and habits. Yet out of this cauldron came Mullis, a young biochemist born in North Carolina. Mullis took a job as a DNA chemist in 1979 at Cetus Corporation in Emeryville, California. A few years later, in 1983, Mullis developed a series of chemical reactions that is now paramount in forensics, genomics, and archaeology.

PCR uses natural enzymes to copy DNA molecules. Cells replicate their DNA when they divide, so that each daughter cell will have its own copies. The enzyme, called DNA polymerase, works by attaching to a single strand of DNA and then catalyzing the reactions that synthesize the complementary strand, which joins the other stand to make a double-stranded helix.

To copy a DNA segment, a solution containing the DNA to be copied, along with plenty of spare nucleotides, is heated so that any double-stranded helices separate (the high temperature breaks the hydrogen bonds holding the two strands together). The DNA polymerase then goes to work. But since the DNA polymerase needs an initial double-stranded segment to get started, PCR technicians include short sequences of DNA in the mix; these short sequences, known as primers, are complementary to the beginning of the segment to be copied. The PCR machine cools the solution slightly so that the primers bind correctly, but not enough for long strands to bind together. After a short period of time, the DNA polymerase finishes, and then the solution is heated again so that the strands separate. The PCR machine repeats this cycle as often as necessary. From even a single piece of DNA, this technique is capable of generating millions of copies in a few hours.

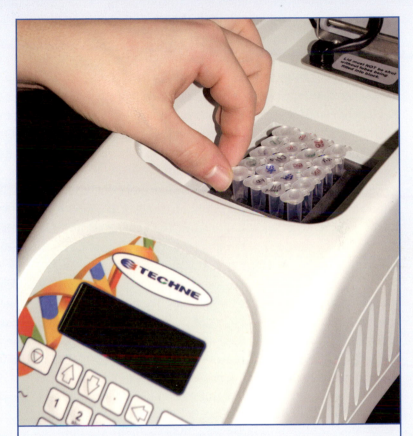

A researcher places tubes of DNA in a PCR machine. *[Martin Shields/Photo Researchers, Inc.]*

The earliest PCR used DNA polymerase from ordinary cells, but the heating caused the enzyme to stop functioning, so a new batch of enzymes had to be added during each cycle. Today, PCR uses a DNA polymerase derived from heat-loving bacteria that live in extremely hot environments. These DNA polymerases are able to maintain their function even at the highest temperature settings in PCR, avoiding the need to replenish the enzymes at each step. Modern PCR machines are automated, requiring the user merely to provide the initial solutions, punch a button, and relax until the operation is complete.

nuclear DNA would have survived, so primers for mitochondrial DNA sequences were chosen.

The operation was a success. As reported in "Molecular Genetic Analyses of the Tyrolean Ice Man," written by Oliva Handt, Svante Pääbo, and many other researchers and published in *Science* in 1994, the researchers found mtDNA fragments that are closely related to people living today in parts of central and northern Europe. Pääbo and his colleagues also reported a lack of success when they used primers for nuclear DNA sequences. Even long mitochondrial sequences were rare; as the researchers noted in their report, "The degradation of the DNA made the enzymatic amplification of mitochondrial DNA fragments of more than 100 to 200 base pairs difficult."

Skeptical observers might wonder if the DNA found in ancient bodies or bones are actually contaminants. PCR amplifies any DNA in the sample, whether it came from the artifact or not. DNA introduced into the sample by humans who handle the artifact is a common headache for ancient DNA research, particularly for bones, which are porous. Bone porosity increases about four or five times after death, which provides a rough field test for archaeologists to gauge the age of a bone—old bones are much more porous. But the increased porosity means that these bones soak up solutions and chemicals, some of which can contain DNA. For the bones of ancient humans, including the Iceman, in which the DNA sequence is expected to be quite closely related to modern humans, it may be difficult to decide if the DNA is ancient or if it is a modern contaminant.

Pääbo and his colleagues took extraordinary precautions in their analyses, comparing their results with controls that did not contain any Iceman samples. The researchers showed that several sequences found in the samples were contaminants, but Ötzi's DNA was slightly different and, as expected, more degraded because of age. As an additional control, the researchers performed several experiments in a different laboratory, showing that the DNA they found was not a contaminant contributed by their own equipment.

Another study of Ötzi's DNA is more controversial. In 2003, the archaeologist Tom Loy at the University of Queensland in Australia tested some of the Iceman's equipment, including his knife, arrows, and coat. Loy had noted signs of blood, and when he used PCR on the samples, he found not one individual's DNA, but four. This evidence suggests that

Ötzi had been in a battle soon before he died, resulting in the blood of several of the combatants staining his clothes and weapons.

Loy's findings are controversial because this work was not published in a science journal. Journalists reported the story based on interviews with Loy, but since contamination is such a major concern with ancient DNA, scientists wanted to learn more about Loy's methods and controls. Loy tragically passed away in 2005 before the details were published. Articles in science journals are important not only in establishing who has done what, but also to let the scientific community examine the procedures and satisfy themselves that everything is in order—and to repeat the experiment, if necessary, to confirm the results. The details need to be published even when a researcher has an excellent reputation because science must be an entirely open and public process if its results are to be accepted without hesitation. In the absence of these details, the validity of Loy's results is not certain, and samples from the Iceman are limited and difficult to obtain.

Using PCR to amplify DNA in fossilized bone has also been used to study samples much older than the Iceman. In 2004, David Serre and Pääbo, now at the Max Planck Institute for Evolutionary Anthropology in Leipzig, Germany, and their colleagues compared mitochondrial DNA found in four Neanderthal fossils from Germany, Russia, and Croatia to that of modern humans. Neanderthals were a species of early humans that lived in parts of Asia and Europe from about 200,000 to 30,000 years ago. Fossil evidence indicates that Neanderthals were a different species than the ancestors of modern humans, though some debate exists over whether Neanderthals may have bred with the ancestors of modern humans, thereby contributing some of their genetic material. Pääbo, Serre, and their colleagues found that the Neanderthal mtDNA sequences were similar to one another but were not found in mtDNA of modern humans. The researchers published their report, "No Evidence of Neandertal mtDNA Contribution to Early Modern Humans," in *PLoS Biology* in 2004. This finding suggests that interbreeding among Neanderthals and the ancestors of modern humans, if it occurred at all, was not significant.

DNA is especially valuable because the sequence can identify a single individual as well as provide a rich source of information about human evolution. But another critical aspect of life is diet—what people eat and drink. Chemical analysis of food and liquids can give archaeologists important clues about how people used to live.

WHAT PHARAOH TUTANKHAMEN DRANK

In 1323 B.C.E., Tutankhamen, the young king—pharaoh—of Egypt, died. In keeping with the traditions of the time, skilled technicians used chemicals to preserve his body for the afterlife. They buried him, along with adequate supplies and a number of precious possessions he would enjoy in his presumed life after death, in an arid valley near the Nile River. There the mummy lay undisturbed—except for perhaps a few hasty and failed attempts by tomb robbers—until the British archaeologist Howard Carter (1874–1939) and his workers dug into the tomb in 1922. Unlike most of Tutankhamen's fellow kings, many of whose tombs lie in the same valley (known as the Valley of the Kings), the tomb of Tutankhamen was relatively intact. Carter and his team uncovered many valuable artifacts, including the pharaoh's mummy.

As pharaoh, Tutankhamen would probably have had the choicest diet. Archaeologists have discovered that kings and the wealthiest Egyptians drank much wine, the production of which from grapes is pictured on tomb walls dating from 4,500 years ago. Wine was important for perhaps a number of reasons, but one of its advantages is that the alcohol tends to kill microorganisms. People in ancient cultures did not know about bacteria, but they did seem to realize that it is a bad idea to drink water that comes from areas where the human population density was high. Without sanitation, waste matter often contaminated water sources around big cities, resulting in bacterial infestation. Not until governments began instituting strict sanitation measures in the late 19th century could safe, clean potable (drinking) water be found in big cities.

Scientists have debated about what type of wine Egyptians drank. Two major classes of wine are red and white (the distinction being in the type of grape, and whether the skin is also used); archaeological chemists have found residues scraped from ancient jars that contain tartaric acid, which is common only in grapes, but have generally been unable to determine the type of wine.

Several jars in Tutankhamen's tomb were labeled "Wine of the House-of-Tutankhamen," although the fluid had evaporated long ago. Maria Rosa Guasch-Jané and her colleagues at the University of Barcelona in Spain recently used sensitive techniques to examine the residues in some of these wine jars. In addition to mass spectrometry, discussed earlier, the researchers used liquid chromatography. This technique separates components of a mixture by pumping a solution to be studied through a column of absorbent material. Some of the components of the solution travel farther along the column than others, and scientists can isolate the separated components. Guasch-Jané and her colleagues discovered a substance, syringic acid, which derives from the grapes used to make red wine. As described in a news release posted on April 3, 2005, at Science*Daily,* this was the first identification of the color of wine in ancient Egyptian residues. In later experiments, the researchers also found evidence for white wine in Tutankhamen's tomb.

Written records of many ancient civilizations, beginning a few thousand years ago, document the use and production of beverages. But archaeological chemists have pushed the date of the earliest such processes further back than the historical records go. A team of researchers led by Patrick E. McGovern, at the Museum Applied Science Center for Archaeology at the University of Pennsylvania Museum of Archaeology and Anthropology, performed chemical analyses of substances that had been absorbed into pottery belonging to Chinese villages of the early stone age. The pottery they studied range from 7,500 to 9,000 years old. Using solvents such as methanol, chloroform, and hexane, along with heat or sonication (high-intensity sound waves), the researchers extracted organic material from the pottery pieces. They identified the materials by a number of different analytical methods, including chromatography, mass spectrometry, and isotope analysis. Components of these materials, such as tartaric acid and alkanes (a type of hydrocarbon), match the composition of wine made from grapes or rice.

The researchers published their findings in a 2004 paper, "Fermented Beverages of Pre- and Proto-historic China," in the *Proceedings of the National Academy of Sciences.* They concluded their paper with the following observation: "The ancient chemical evidence now enables the later beverages to be traced back as far as 7000 B.C.E. and reveals how Chinese beverage-making developed over the millennia. Our results also illustrate how both religious ceremonies and activities of everyday life in which these vessels were used, and still important in modern Chinese culture, likely have their basis in prehistory."

The Museum Applied Science Center for Archaeology conducts studies such as these to expand the knowledge of ancient cultures by applying rigorous scientific techniques. More information about this research center can be found in the following sidebar.

Museum Applied Science Center for Archaeology

Museums such as the Smithsonian Institution in Washington, D.C., and many other museums across the globe, display a variety of artifacts from past cultures. Tools, weapons, utensils, fabrics, and bones provide a rich visual description of how people used to live and die. By visiting a museum, a person can take a step back in time and get a sense of the problems faced by people who came before, as well as the tools and technology by which these people attempted to solve these problems.

Yet there is even more knowledge to be gleaned from these artifacts. Organizations such as the Museum Applied Science Center for Archaeology, which is the scientific branch of the University of Pennsylvania Museum of Archaeology and Anthropology, strives to analyze and interpret excavated artifacts with the aid of science and technology. This museum, located on the campus of the University of Pennsylvania in Philadelphia, opened in 1899 and features artifacts from ancient Egypt, Mesopotamia, Asia, and much more.

Artifacts at the University of Pennsylvania's museum provide a rich source of material to explore. Scientific studies at the Museum Applied Science Center for Archaeology include archaeometallurgy (the study of metal tools and weapons), faunal analysis (the study of animal bones), and pottery analysis. The staff also performs a lot of organics analysis, such as the wine experiments described in the text. In 2007, the research center participated in a study that showed the earliest known use of chocolate—a chocolate beverage made from cacao beans around 1400–1100 B.C.E. in Honduras, as discovered from an analysis of pottery residues. John S. Henderson at Cornell University, along with researchers at the Museum Applied Science Center for Archaeology, published this result in "Chemical and Archaeological Evidence for the Earliest Cacao Beverages," in the *Proceedings of the National Academy of Sciences*.

Analyses of food and drink are not limited to the residues of unconsumed substances. In the Iceman's case, scientists had access to the body—and therefore to Ötzi's last ingested meals.

Franco Rollo and his colleagues at the University of Camerino in Italy studied DNA extracted from samples of Ötzi's intestinal contents. Using PCR, with primers designed to amplify a number of different genes from various plants and animals, the researchers discovered the remains of two meals, based on their position in the intestines. (Ötzi's stomach was empty, which meant that he had not eaten in a few hours before the time of death—the meals had enough time to move into the intestines.) One sample came from the ileum (the final portion of the small intestine), and the other, which had moved farther down the digestive system, came from the colon in the large intestine. The earlier meal, the remains of which were in the colon, contained DNA from an ibex (a mountain goat) and some cereal plants. Ötzi's last meal, located in the ileum, consisted of red deer and cereals. The researchers published their report, "Ötzi's Last Meals: DNA Analysis of the Intestinal Content of the Neolithic Glacier Mummy from the Alps," in a 2002 issue of the *Proceedings of the National Academy of Sciences*.

REVISITING THE PAST

The chemical techniques discussed in this chapter have granted archaeologists an incredible ability to revisit the past. With the aid of chemistry, archaeologists now believe they have a rough idea of Ötzi's last days. This knowledge includes the analysis of his meals but goes much further—researchers have used the intestinal contents to track Ötzi's final journey. The key component in this study was pollen.

Pollen consists of reproductive cells released by plants (and sometimes carried by insects) at certain times of the year. Individual species of plant are identifiable by the grains of pollen they generate. Klaus Oeggl of Innsbruck University, James H. Dickson at the University of Glasgow in Scotland, and their colleagues have studied pollen samples obtained from Ötzi's digestive tract. Some of this pollen may have been eaten intentionally, but Ötzi probably swallowed most of it accidentally, either in the course of eating a meal or by inhalation.

Oeggl and his colleagues discovered pollen from the hop hornbeam tree, which grows in warm environments at lower altitudes. This tree blooms in the springtime, indicating that the Iceman died in spring.

This is surprising since conditions would still be harsh at this time in the mountains, making the trip he was taking when he died quite perilous. Above and below the hop hornbeam pollen, the researchers found traces of pine pollen in Ötzi's digestive tract. Pine grows in higher elevations. What this indicates is that Ötzi was in the mountains for a while, then descended—probably to a village—and then quickly climbed again, where he died at 10,530 feet (3,210 m) above sea level. The researchers published their report, "The Reconstruction of the Last Itinerary of 'Ötzi', the Neolithic Iceman, by Pollen Analyses from Sequentially Sampled Gut Extracts," in *Quaternary Science Reviews* in 2007.

In 2005, scientists discovered the cause of Ötzi's death. A few years earlier, in 2001, Paul Gostner wheeled an X-ray machine to Ötzi's chamber in the South Tyrol Museum. Gostner, who works at the Central Hospital in Bolzano, Italy, discovered a stone arrowhead embedded in the back of Iceman's left shoulder. Earlier X-ray scans had missed the small object. The arrow shaft was not in Ötzi's body, nor had it been found nearby. The fatality of the wound became evident in 2005, when Central Hospital acquired an X-ray machine with higher resolution. Gostner and other staff members of the hospital brought Ötzi in for a scan—it was a rush job, otherwise Ötzi's body would quickly decompose. They discovered that the arrowhead had gashed a large and important artery, the subclavian artery, which carries blood to the arm. Such a serious injury would have caused Ötzi to bleed to death in minutes.

The arrowhead was yet another important clue in archaeologists' attempts to reconstruct the Iceman's final days. Somebody apparently wanted Ötzi dead. The killer was a good shot and had no qualms about shooting a person from behind. Perhaps Loy's controversial DNA findings, which indicated Ötzi's involvement in a recent fight, are true. Supporting this idea, one of Ötzi's thumbs had been severely cut, which must have occurred soon before he died since the wound had not healed. Ötzi may have been fleeing a pursuer, which would explain what he was doing in the mountains in the hazardous and snowy springtime.

One theory, based on the pollen analysis, suggests that when Ötzi returned from a trip to higher altitudes, he got into a dispute with some people in the village. According to the isotope evidence coming from the Iceman's bones and teeth, he was familiar with the area, so he would have probably known the villagers—they may have even been relatives. Ötzi retreated back into the mountains after the fight, only to die at the hands of a skilled archer.

But if this scenario is true, why was Ötzi in the mountains in the first place? Perhaps the fight occurred earlier, and Ötzi fled into the mountains. Then, in accordance with the pollen analysis, he descended briefly, maybe to retrieve some supplies. But this descent may have been his undoing, for he might have been seen and tracked. High in the mountains, where no one would learn of the deed—at least not until 5,300 years later—the killer struck.

Much uncertainty remains. But as archaeological chemists develop increasingly sophisticated analyses, further clues may emerge about the Iceman's life and death.

THE FALL OF ROME

The tales of individuals such as Ötzi reveal something about the culture in which they lived. In past times, as in the present, disputes have a way of escalating into extreme violence. Other archaeological chemistry concerns the fate of empires.

The Romans dominated the region surrounding the Mediterranean Sea, including much of Europe and the Near East, for hundreds of years. Named after Rome, the capital, which was founded in the seventh or sixth century B.C.E., the Romans became the dominant power in the region after defeating the Carthaginians in a series of wars in the third and second centuries B.C.E. This civilization greatly influenced the politics, language, religion, and justice system for most of Europe and subsequently the United States all to the way to the present day.

Yet somehow Rome fell to invading "barbarians"—the name Romans applied to foreigners—in the fifth century C.E. Although the eastern branch of the Roman Empire, the Byzantine Empire, survived for another thousand years, the mighty Roman civilization around the Mediterranean Sea collapsed. Historians have been debating who or what was responsible for the collapse ever since.

Theories abound. The British historian Edward Gibbon (1737–94), whose multivolume work, *The History of the Decline and Fall of the Roman Empire*, published in 1776–88, blamed the collapse on overindulgence, lapses in military strength, and a weakening of the people's respect for their leaders following the Roman government's adoption of Christianity in the empire's later stages.

It is unlikely that any single event or policy was solely responsible for the fall. With the vast and intricate political and social system of the widespread Romans, several factors probably contributed to the weakness that was eventually exposed by the "barbarians." Archaeological chemistry may be able to help decide which factors were the most important.

One possibility involves the use of lead. Although heavy, this abundant metal is easily worked—lead is soft, malleable, and has a low melting point. The Romans fashioned water pipes out of lead and used them in many areas of their empire, especially Rome. So common was the use of lead for this purpose that it influenced the English term *plumbing,* which derives from the Latin word for lead, *plumbum.* (This Latin word also explains the chemical symbol for lead, which is Pb.) Romans also commonly boiled fruits such as grapes in lead vessels.

The use of lead in Roman and Greek civilizations was so great that it caused widespread air pollution, which spread around the globe. Some of this lead settled, along with snow, onto ice packs in Greenland, where it lay on the surface until subsequent layers buried it. The snow turned to ice as layers accumulated undisturbed over the millennia. At each depth of the pack, the composition of the ice provides information on climate and atmospheric composition at the time of the deposit. Researchers can procure an ice core by drilling into the ice with a hollow bit, then chemically analyze the ice at various depths to study the atmosphere in ancient times. In 1994, the geologists Sungmin Hong, Claude F. Boutron, and their colleagues analyzed an ice core from Greenland that covered a period from 3,000 to 500 years ago. As reported in "Greenland Ice Evidence of Hemispheric Lead Pollution Two Millennia Ago by Greek and Roman Civilizations," published in *Science,* they found the concentration of lead was four times higher than normal during the Roman era.

But lead is poisonous. Lead poisoning can cause elevated blood pressure, pain, mental disorders, irritability, and infertility. The Romans were not completely unaware of the dangers of lead, for they noted the poor health of the unfortunate men who produced and worked with the metal. Yet this warning may not have been enough. Perhaps the Romans believed that their exposure to lead leaching from pipes or cooking vessels would not be serious.

Did the plentiful use of lead cause severe outbreaks of lead poisoning in the Romans? Archaeological chemists looking into this issue have examined skeletons dating from Roman times. Lead accumulates in the bones as the body absorbs this heavy element. The University of Minnesota researcher Arthur C. Aufderheide and his colleagues tested Roman bones for lead, discovering in 1992 that skeletons dating to later periods of Roman history have 10 times more lead than their predecessors.

But as with DNA, the question of contamination arises. Because of the porosity of old bones, they can absorb substances from the soil in which they lie. Since lead is a common component of the soil, skeptical archaeologists may wonder if the lead in Roman skeletons is coming from the soil, absorbed after death, rather than being due to exposure during life.

Researchers are addressing this issue by measuring isotopes. Lead in the soil often has a different ratio of isotopes from the lead that was absorbed by the body and subsequently deposited in the bones. David De Muynck, a researcher at Ghent University in Belgium, and his colleagues studied 22 samples of infant bone tissue dating from Roman times. These bones had high concentrations of lead. By comparing lead isotope ratios of the bones with that in the soil and other objects in the graves, the researchers determined that bone absorption after death did not contribute most of the lead in the bones. The report, "Lead Isotopic Analysis of Infant Bone Tissue Dating from the Roman Era Via Multicollector ICP-Mass Spectrometry," was published in a 2008 issue of *Analytical and Bioanalytical Chemistry.*

Although this and other research suggests the Romans were exposed to a considerable amount of lead—and they absorbed quite a bit of it—no one yet knows if they were adversely affected. Lead poisoning could explain why Roman children suffered from high mortality; as De Muynck and his colleagues noted in their paper, "Approximately 26% of Roman children died before the age of fourteen, while approximately 14% even died in the first year of life." The disorder may also account for the apparent madness of emperors such as Caligula and Nero. However, these and other problems could easily be due to other causes.

Archaeological chemistry is just beginning to explore the life and death of ancient peoples and civilizations. This research opens a vast avenue to reach back into history and revisit the past. Further innovations, along with the ingenuity of researchers, will grant archaeologists an even better view of times long ago.

CONCLUSION

Historians and archaeologists explore the past by gathering artifacts and using technology to glean as much knowledge as they can from them.

Some imaginative writers have described scenarios in which researchers not only study the past, they recreate it. In Michael Crichton's 1990 novel *Jurassic Park*, scientists establish a dinosaur park that resembles the Jurassic era by recreating dinosaurs based on ancient DNA sources.

The novel, and subsequent film, offered an exciting story, but one that is unlikely to ever become a reality. DNA degrades, even DNA deposited in locations with little chemical activity. For instance, although parts of Ötzi's body were preserved for more than 5,000 years, researchers could not recover any nuclear DNA and found only bits and pieces of mitochondrial DNA. Since the dinosaurs became extinct about 65 million years ago, any hope of recovering enough dinosaur DNA to duplicate or reproduce these animals is unrealistic.

Yet it is not impossible to gather enough clues to find out what happened long ago. Just as paleobiologists—scientists who study ancient life—have theorized about a comet or asteroid impact causing the dinosaur extinction, archaeological chemists have studied artifacts to learn something about the life and death of Ötzi, as well as the rise and fall of Roman civilization.

As research on these projects continues, other projects are getting started. Zhichun Jing, a researcher at the University of British Columbia in Canada, has an ambitious plan to study Chinese civilization to find out why societies rise and decline. China has witnessed thousands of years of settlements, some rising to great heights in terms of technological achievement and political stability, such as the Shang dynasty of 1200 to 1050 B.C.E., then falling into disarray. Supported by several funding agencies in Canada and the United States, this research could soon yield a better understanding of the dynamics of social and political systems.

Many of the problems facing the United States and the rest of the world are probably similar to those encountered by the successful civilizations of the past. Environmental pollution, increasing population, scarcity of resources, as well as disruptions caused by weather and climate, pose significant obstacles to the health of the economy and the high standards of living. Solutions to these problems may lie in the development of new and sophisticated technology, but the dilemmas faced by people of the past, and their successes and failures, could also play an important role in the critical decisions that people of today and tomorrow must make.

CHRONOLOGY

1784 C.E. American statesman and scientist Thomas Jefferson (1743–1826) directs one of the earliest archaeological investigations—an excavation of a Native American mound in Virginia. This and other early excavations help to establish the scientific methods of archaeology.

1903 Russian scientist Mikhail Tswett (1872–1919) publishes the first description of chromatography.

1913 American chemist Theodore Richards (1868–1928) discovers lead of varying mass (due to different isotopes).

British chemist Frederick Soddy (1877–1956) develops the concept of isotopes.

1919 Building upon the work of others, British scientist Francis Aston (1877–1945) develops the mass spectrometer.

1922 British archaeologist Howard Carter (1874–1939) leads the exploration of Egyptian pharaoh Tutankhamen's tomb.

1949 University of Chicago chemist Willard Libby (1908–80) develops the technique of radiocarbon dating.

1960s British scientist Francis Crick (1916–2004), South African biologist Sidney Brenner (1927–   ), and other researchers establish the genetic code by which the sequence of DNA can be understood. The development of advanced sequencing techniques, as used in many branches of biology as well as archaeological chemistry, followed.

1983	Kary Mullis, a chemist working for Cetus Corporation, develops polymerase chain reaction, a technique subsequently used to amplify and detect tiny amounts of DNA such as that available from ancient sources.
	Environmental chemist Jerome Nriagu publishes *Lead and Lead Poisoning in Antiquity,* in which he argues that lead poisoning contributed to the decline and fall of the Roman Empire. Although the idea had been discussed earlier, Nriagu's argument reopened the debate with considerable intensity.
1991	German tourists hiking in the Italian Alps discover Ötzi the Iceman, a remarkably well preserved 5,300-year-old body.
1994	Geologist Claude F. Boutron and colleagues discovers high amounts of lead in Greenland ice cores dating from Roman times, confirming widespread production and use of lead in this period.
2002	Franco Rollo and colleagues determine the composition of Ötzi's last meals by examining the DNA sequences of material found in his intestines.
2003	Wolfgang Müller and his colleagues use isotope measurements of Ötzi's teeth and bones to suggest the geographical locations of his childhood and recent habitat.
2005	Paul Gostner and other physicians make high-resolution X-ray scans of Ötzi's shoulder, revealing that the arrow wound gashed an important artery that would have resulted in his quick death.
2007	David De Muynck and colleagues determine that high concentrations of lead in the bones of a group of Roman infants were not primarily due to absorption after death, suggesting that considerable absorption occurred during the individuals' lifetime.

FURTHER RESOURCES
Print and Internet

Bahn, Paul. *Archaeology: A Very Short Introduction.* Oxford: Oxford University Press, 2000. There are many texts on archaeology, but this book, as the title implies, covers the ground (so to speak) with brevity. But all the major topics, including the techniques of archaeology, are discussed in these 128 pages.

Cowell, F. R. *Life in Ancient Rome.* New York: Perigee Books, 1980. This reprint of a 1961 book describes in eloquent detail how ancient Romans raised their families, earned a living, spent their leisure time, and worshipped their gods.

Deem, James M. "Mummy Tombs." Available online. URL: http://www.mummytombs.com/. Accessed May 28, 2009. A splendid educational resource, these pages contains information on Egyptian mummies, Ötzi, and the latest mummy news.

De Muynck, David, Christophe Cloquet, Elisabeth Smits, Frederik A. de Wolff, Ghylaine Quitté, Luc Moens, and Frank Vanhaecke. "Lead Isotopic Analysis of Infant Bone Tissue Dating from the Roman Era Via Multicollector ICP-Mass Spectrometry." *Analytical and Bioanalytical Chemistry* 390 (2008): 477–486. The researchers used isotope analysis to show that high concentrations of lead in the bones of Roman infants probably did not come from the soil or other objects in the graves.

Feder, Kenneth L. *Frauds, Myths, and Mysteries: Science and Pseudoscience in Archaeology,* 5th ed. New York: McGraw-Hill, 2005. Archaeology, as with any science, has suffered from highly publicized claims backed with little or faulty scientific evidence. In describing some of these cases, such as the Piltdown hoax, this book discusses the scientific techniques and procedures that separate valid findings from unlikely claims.

Fowler, Brenda. *Iceman: Uncovering the Life and Times of a Prehistoric Man Found in an Alpine Glacier.* New York: Random House, 2000. Fowler, a journalist, chronicles the discovery and subsequent studies of Ötzi the Iceman. Although a lot of research has occurred since the publication of this book, the author provides a wealth of interesting detail on the people, politics, and science associated with the period shortly following this amazing discovery.

Handt, Oliva, Martin Richards, Marion Trommsdorff, Christian Kilger, Jaana Simanainen, Oleg Georgiev, et al. "Molecular Genetic Analyses of the Tyrolean Ice Man." *Science* 264 (June 17, 1994): 1,775–1,778. The researchers found genetic fragments in the Iceman's remains that are closely related to people living today in parts of central and northern Europe.

Henderson, John S., Rosemary A. Joyce, Gretchen R. Hall, W. Jeffrey Hurst, and Patrick E. McGovern. "Chemical and Archaeological Evidence for the Earliest Cacao Beverages." *Proceedings of the National Academy of Sciences* 104 (November 27, 2007): 18,937–18,940. The researchers studied pottery residues and found that a chocolate beverage had been made from cacao beans in 1400–1100 B.C.E. in Honduras.

Hong, Sungmin, Jean-Pierre Candelone, Clair C. Patterson, and Claude F. Boutron. "Greenland Ice Evidence of Hemispheric Lead Pollution Two Millennia Ago by Greek and Roman Civilizations." *Science* 265 (September 23, 1994): 1,841–1,843. The researchers analyzed an ice core from Greenland that covered a period from 3,000 to 500 years ago and found the concentration of lead was four times higher than natural conditions during the Roman era.

Lambert, Joseph B. *Traces of the Past: Unraveling the Secrets of Archaeology through Chemistry.* Cambridge, Mass.: Perseus Publishing, 1997. Lambert, a chemistry professor at Northwestern University, explains the chemical methods of archaeology and how they help reconstruct the past. Chemical analyses of stone, pottery, glass, pigments, metals, and biological materials are discussed.

McGovern, Patrick E., Juzhong Zhang, Jigen Tang, Zhiqing Zhang, Gretchen R. Hall, Robert A. Moreau, et al. "Fermented Beverages of Pre- and Proto-historic China." *Proceedings of the National Academy of Sciences* 101 (December 21, 2004): 17,593–17,598. The researchers use chemical techniques on archaeological samples to explore how ancient Chinese developed fermented beverages.

Müller, Wolfgang, Henry Fricke, Alex N. Halliday, Malcolm T. McCulloch, and Jo-Anne Wartho. "Origin and Migration of the Alpine Iceman." *Science* 302 (October 31, 2003): 862–866. The researchers performed isotope measurements of Ötzi's teeth and bones and compared these values with surrounding soils, rocks, and streams.

Oeggl, Klaus, Werner Kofler, Alexandra Schmidl, James H. Dickson, Eduard Egarter-Vigl, and Othmar Gaber. "The Reconstruction of the Last Itinerary of 'Ötzi', the Neolithic Iceman, by Pollen Analyses from Sequentially Sampled Gut Extracts." *Quaternary Science Reviews* 26 (2007): 853–861. The researchers found pollen samples in Ötzi's digestive tract that indicates Ötzi was in the mountains for a while, then descended—probably to a village—and then quickly climbed again to the place where he died.

Perre, David, André Langaney, Mario Chech, Maria Teschler-Nicola, Maja Paunovic, Philippe Mennecier, et al. "No Evidence of Neandertal mtDNA Contribution to Early Modern Humans." *PLoS Biology.* March 2004. Available online. URL: http://www.plosbiology.org/article/info: doi/10.1371/journal.pbio.0020057. Accessed May 28, 2009. The researchers compared mitochondrial DNA found in four Neanderthal fossils from Germany, Russia, and Croatia to that of modern humans. They found that the Neanderthal mtDNA sequences were similar to one another but were not found in mtDNA of modern humans.

Rollo, Franco, Massimo Ubaldi, Luca Ermini, and Isolina Marota. "Ötzi's Last Meals: DNA Analysis of the Intestinal Content of the Neolithic Glacier Mummy from the Alps." *Proceedings of the National Academy of Sciences* 99 (October 1, 2002): 12,594–12,599. The researchers discovered the remains of two meals, based on their position in the intestines.

Science*Daily.* "King Tut Liked Red Wine." News release, April 3, 2005. Available online. URL: http://www.sciencedaily.com/releases/2005/03/050326001121.htm. Accessed May 28, 2009. Researchers identify the color of wine in ancient samples.

Web Sites

Archaeological Institute of America: The WWWorld of Archaeology. Available online. URL: http://www.archaeology.org/wwwarky/. Accessed May 28, 2009. This Web site contains links to interesting sites about archaeology. Categories include classical archaeology (Greek and Roman archaeology), Europe, North America, Asia and the Pacific Ocean, Africa and Egypt, underwater archaeology, and others.

Cold Spring Harbor: DNA Interactive. Available online. URL: http://www.dnai.org/. Accessed May 28, 2009. Graphics and animations enrich the presentation of this Web site, which describes DNA, genetics, and the history and development of DNA research. There is also a section describing applications, such as the use of DNA in archaeological investigations.

Illustrated History of the Roman Empire. Available online. URL: http://www.roman-empire.net/. Accessed May 28, 2009. This Web site contains an enormous collection of information on Roman civilization. Chapters describe the various periods of Roman history, such as the early republic, the expanding empire, and its decline and fall. Maps, lists of emperors and battles, and photographs of the ruins are included.

Museum Applied Science Center for Archaeology. Available online. URL: http://masca.museum.upenn.edu/. Accessed May 28, 2009. As the scientific branch of the University of Pennsylvania Museum of Archaeology and Anthropology, the Museum Applied Science Center for Archaeology conducts research on many topics, including archaeological chemistry. Their Web site describes the techniques they use and some of their current projects.

South Tyrol Museum of Archaeology: Oetzi/Ötzi, the Iceman. Available online. URL: http://www.archaeologiemuseum.it/f01_ice_uk.html. Accessed May 28, 2009. The Web site of the South Tyrol Museum of Archaeology, which houses Ötzi, presents photographs and information on the mummy, clothing, and equipment.

FINAL THOUGHTS

Much of the chemistry described in this book involves research on how elements combine and how to control or manipulate these reactions to extract energy, fashion miniature motors, or produce medications that act in the brain to reduce depression and other disorders. Another important aspect of chemistry focuses on the elements themselves—and how to make new ones.

This field of research has a certain similarity to the old ideas of alchemy, in which people sought the means to manufacture precious elements such as gold. But unlike the alchemists, whose concoctions and recipes failed to deliver the desired substance, modern chemists have succeeded in generating new elements as well as new isotopes of old elements. All elements with an atomic number higher than uranium were initially discovered in the laboratory and are never or only rarely found in nature. Uranium's atomic number is 92, and the first heavier element to be discovered was neptunium, which Edwin McMillan (1907–91) and Philip Abelson (1913–2004) found while conducting nuclear research at Berkeley Radiation Laboratory in California in 1940. Neptunium's atomic number is 93. The elements having an atomic number greater than uranium are collectively known as the transuranium elements. (Neptunium was named after the planet Neptune. The logic behind the name was that neptunium follows uranium in the sequence of elements, and Neptune follows Uranus in the sequence of planets when listed in order of increasing distance from the Sun.)

Scientists have found about 270 different nuclides—types of nuclei—but roughly 3,000 have been fashioned in the laboratory. (The term *isotope* refers to different forms of the same element—for example, carbon 12, carbon 13, and carbon 14 are isotopes of carbon. The term *nuclide* is more general, referring to different forms of any or all the elements, such as carbon 14, oxygen 16, uranium 238, and so on.) Many of these nuclides, particularly the new ones, decay. This decay, known as radioactive decay, occurs when one nucleus transforms into another, emitting radiation (radioactivity) in the process. For example, carbon 14 decays to nitrogen 14, as discussed in the sidebar "Radiocarbon Dating" on page 170. Atoms that are stable, such as carbon 12, do not decay.

Despite their instability, some unstable atoms may last a long time; the half-life of uranium 238, for example, is about 4.5 billion years. Other unstable atoms decay in a few seconds. Radioactive decay is one of the topics of nuclear chemistry, and it involves nuclear forces, as governed by advanced concepts in chemistry and physics, such as quantum mechanics. Researchers do not fully understand why some atoms are stable and others are not, but most radioactive nuclei have an unusually large (or small) number of neutrons, which makes the nucleus unstable. And all heavy nuclei found so far are radioactive—nuclides with an atomic number of 83 or greater decay.

Smaller elements found in nature, such as hydrogen and helium, formed shortly after the birth of the universe, some 14 billion years ago. Heavier elements such as oxygen, iron, and gold formed in the nuclear reactions of stars such as the Sun during their lifetimes or, in the case of the heaviest elements, in nuclear reactions that occur at the end of a large star's lifetime, when it explodes and becomes what astronomers call a supernova.

As evidenced by the tremendous power of nuclear bombs, nuclear reactions involve quite a lot of energy. In the laboratory, researchers fabricate nuclides with the aid of special, high-energy equipment such as reactors in which nuclear reactions can take place, or particle accelerators in which particles such as protons are accelerated to high speed and crash into one another, or some other target. For example, in 2006, researchers at the Joint Institute for Nuclear Research in the Russian Federation and the Lawrence Livermore National Laboratory in California synthesized isotopes of element 118 for the first time. To make the new isotope, researchers smashed calcium atoms into a target made of californium (which has an atomic number of 98). These new isotopes decayed quickly. (Element 118 and other recently discovered elements have not yet been named.)

Studying these isotopes provides fertile ground for physicists and chemists to gain a better understanding of the properties and behavior of nuclei. This field of research also has important applications. For example, radioisotopes—radioactive isotopes—that emit certain particles are critical in some medical treatments such as radiation therapy, which is used to kill cancer cells, and positron emission tomography (PET), which is extremely useful in imaging parts of the body.

Making new nuclides is a difficult, energy-intensive process, and no one knows how many more of them can be made, or what sort of properties they may have. But researchers such as Bradley M. Sherrill at Michigan State University advocate more research, likening the process to nanotechnology, at the atomic scale—the production of "designer" atomic nuclei. Potential applications include medical treatments that might be able to use new isotopes to fight diseases such as cancer, as well as nuclear energy utilities, which could benefit from new and cleaner nuclear reactions.

Chemists have already learned much by studying the isotopes found in nature, as well as those that have been created in the laboratory. An increase in the number of available atoms means an increase in the "raw material" of chemistry, which will give future chemists even greater opportunities to expand the frontiers of their science.

APPENDIX A

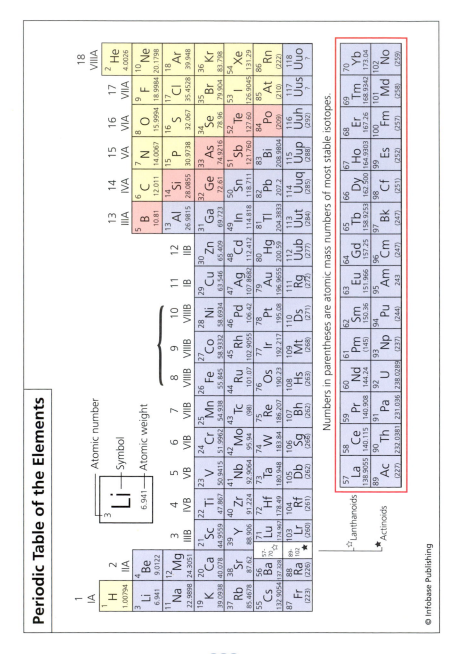

Periodic Table of the Elements

Numbers in parentheses are atomic mass numbers of most stable isotopes.

☆ Lanthanoids

★ Actinoids

© Infobase Publishing

The Chemical Elements

(g) none (c) metallics

element	symbol	a.n.
aluminum	Al	13
bohrium	Bh	107
cadmium	Cd	48
chromium	Cr	24
cobalt	Co	27
copper	Cu***	29
darmstadtium	Ds	110
dubnium	Db	105
gallium	Ga	31
gold	Au***	79
hafnium	Hf	72
hassium	Hs	108
indium	In	49
iridium	Ir ****	77
iron	Fe	26
lawrencium	Lr	103
lead	Pb	82
lutetium	Lu	71
manganese	Mn	25
meitnerium	Mt	109
mercury	Hg	80
molybdenum	Mo	42
nickel	Ni	28
niobium	Nb	41
osmium	Os****	76
palladium	Pd****	46
platinum	Pt ****	78
rhenium	Re	75
rhodium	Rh****	45
roentgenium	Rg	111
ruthenium	Ru****	44
rutherfordium	Rf	104

(g) none (c) metallics

element	symbol	a.n.
scandium	Sc	21
seaborgium	Sg	106
silver	Ag***	47
tantalum	Ta	73
technetium	Tc	43
thallium	Tl	81
titanium	Ti	22
tin	Sn	50
tungsten	W	74
ununbium	Uub	112
ununtrium	Uut	113
ununquadium	Uuq	114
vanadium	V	23
yttrium	Y	39
zinc	Zn	30
zirconium	Zr	40

(g) pnictogen (c) metallics

element	symbol	a.n.
arsenic	As*	33
antimony	Sb*	51
bismuth	Bi	83
nitrogen	N	7
phosphorus	P**	15
ununpentium	Uup	115

(g) none (c) semimetallics

element	symbol	a.n.
boron	B	5
germanium	Ge	32
silicon	Si	14

(g) lanthanoid (c) metallics

element	symbol	a.n.
cerium	Ce	58
dysprosium	Dy	66
erbium	Er	68
europium	Eu	63
gadolinium	Gd	64
holmium	Ho	67
lanthanum	La	57
neodymium	Nd	60
praseodymium	Pr	59
promethium	Pm	61
samarium	Sm	62
terbium	Tb	65
thulium	Tm	69
ytterbium	Yb	70

(g) actinoid (c) metallics

element	symbol	a.n.
actinium	Ac	89
americium	Am	95
berkelium	Bk	97
californium	Cf	98
curium	Cm	96
einsteinium	Es	99
fermium	Fm	100
mendelevium	Md	101
neptunium	Np	93
nobelium	No	102
plutonium	Pu	94
protactinium	Pa	91
thorium	Th	90
uranium	U	92

(g) noble gases (c) nonmetallics

element	symbol	a.n.
argon	Ar	18
helium	He	2
krypton	Kr	36
neon	Ne	10
radon	Rn	86
xenon	Xe	54
ununoctium	Uuo	118

(g) halogens (c) nonmetallics

element	symbol	a.n.
astatine	At*	85
bromine	Br	35
chlorine	Cl	17
fluorine	F	9
iodine	I	53
ununseptium	Uus*	117

(g) none (c) nonmetallics

element	symbol	a.n.
carbon	C	6
hydrogen	H	1

(g) chalcogen (c) nonmetallics

element	symbol	a.n.
oxygen	O	8
polonium	Po	84
selenium	Se	34
sulfur	S	16
tellurium	Te	52
ununhexium	Uuh	116

(g) alkali metal (c) metallics

element	symbol	a.n.
cesium	Cs	55
francium	Fr	87
lithium	Li	3
potassium	K	19
rubidium	Rb	37
sodium	Na	11

(g) alkaline earth metal (c) metallics

element	symbol	a.n.
barium	Ba	56
beryllium	Be	4
calcium	Ca	20
magnesium	Mg	12
radium	Ra	88
strontium	Sr	38

* = semimetallics (c)
** = nonmetallics (c)
*** = coinage metal (g)
**** = precious metal (g)

a.n. = atomic number
(g) = group
(c) = classification

GLOSSARY

action potentials   electrical impulses, lasting only a few milliseconds, generated in neurons and other electrically active biological cells

alloys   mixtures of two or more elements, at least one of which is generally a metal

artifacts   human-made or human-modified objects

atomic number   the number of protons in an atom's nucleus, which specifies the element

atoms   the smallest particle of an element, generally consisting of electrons "orbiting" a nucleus containing protons and neutrons

cancer   any of a number of diseases marked by abnormal growth of cells and tissues

ceramic   a material containing a nonmetallic mineral, such as clay, baked at high temperature

chromatography   technique used to separate mixtures into their components based on differences in mobility

composites   a material made of small fibers of one substance embedded in a matrix of another substance

compounds   substances consisting of elements combined in fixed proportions

covalent bond   strong chemical interaction between atoms and molecules formed by electron sharing

crystal a solid having a regular, repeating arrangement of components

deoxyribonucleic acid substance consisting of units called bases, the sequence of which represents an organism's genetic information

distillation technique used to separate mixtures based on differences in boiling points

DNA *See* **deoxyribonucleic acid**

electrodes materials that conduct electrical charges

electrolyte a substance, often a fluid, containing mobile electrical charges such as ions that are capable of carrying a current

electrons negatively charged particle that "orbits" an atom's nucleus, or, in the free state, may carry electrical current

elements fundamental chemical substances that cannot be broken down into simpler chemical substances

half-life the amount of time in which one half of a radioactive substance will decay

ionic bond strong chemical interaction between atoms and molecules formed by electron transfer

ions electrically charged particles

isotopes different forms of the same element, which vary in the number of neutrons

laser a device capable of generating an intense beam of light or other electromagnetic radiation

macroscopic large enough to be visible with the unaided eye, as opposed to microscopic, seen only through a microscope

magnetostriction change in shape of a material in response to the application of a magnetic field

metabolism all the chemical reactions through which a biological cell or organism sustains life

metabolites   products of certain metabolic activity

molecules   units of two or more atoms held together with chemical bonds

nanometer   unit of distance equal to a billionth of a meter, or 0.00000004 inches (0.0000001 cm)

nanoparticles   particles of the size typically encountered in nanotechnology—0.00000004–0.000004 inches (1–100 nm)

neuron   brain cells whose electrochemical activity processes information

neurotransmitter   molecule released by neurons for the purpose of signaling or affecting other neurons or other biological cells

neutrons   electrically neutral particles usually found in the nucleus of an atom

NMR   *See* **nuclear magnetic resonance**

nuclear magnetic resonance   certain properties displayed by substances exposed to a magnetic field, from which characteristics of the substance, such as its molecular structure, can be studied

nucleus   central portion of an atom, containing protons and neutrons

organic   substances having a carbon composition, and usually associated with living organisms

oxidation-reduction   a chemical reaction in which atoms gain or lose, or appear to gain or lose, electrons

piezoelectric   capable of generating electricity in response to mechanical stress

proteins   large molecules that perform a variety of functions in biological tissues, and are composed of a sequence of smaller molecules known as amino acids

protons   positively charged particles usually found in the nucleus of an atom

quantum dot   tiny semiconductor that exhibits properties distinct from large, bulky semiconductors

quantum mechanics   a set of concepts and equations in advanced chemistry and physics that describe the behavior of small particles such as atoms and their components

reactions   chemical processes in which one or more substances are transformed

receptor   in brain chemistry, a particle or molecule, usually a protein, in or on a biological cell, to which signaling molecules such as neurotransmitters bind and exert an effect

semiconductor   a material that conducts electricity under certain conditions

solvent   a substance that dissolves other substances

synapse   junction between two neurons through which the neurons transfer information

thermodynamics   the study of heat and related phenomena

X-ray crystallography   technique using the scattering of X-rays to determine the atomic structure of a crystal

X-rays   electromagnetic radiation with a much higher frequency (and therefore smaller wavelength) than visible light

FURTHER RESOURCES

Print and Internet

Bentor, Yinon. "Chemical Elements." Available online. URL: http://www. chemicalelements.com/. Accessed May 28, 2009. Created as an 8th grade science project, this Web site has a clickable periodic table of elements. Clicking the element symbol leads to a page containing basic information, atomic structure, a list of isotopes, and several other facts about that element.

Cobb, Cathy, and Harold Goldwhite. *Creations of Fire: Chemistry's Lively History from Alchemy to the Atomic Age.* New York: Basic Books, 2002. The authors, both chemists, describe the development of chemical knowledge—from the confusion of alchemists who tried to turn lead into gold to the onset of nuclear and quantum chemistry, where people learned to tame and harness the power of the atom.

Cobb, Cathy, and Monty L. Fetterolf. *The Joy of Chemistry: The Amazing Science of Familiar Things.* Amherst, N.Y.: Prometheus Books, 2005. The authors approach the subject of chemistry by relating how its scientific principles influence human activity, society, and technology. Respiration, diamonds, refrigerators, solar energy, aspirin, and many other topics are included.

Emsley, John. *Nature's Building Blocks: An A-Z Guide to the Elements.* Oxford: Oxford University Press, 2003. Each element has an entry. The entry provides some basic information, such as the element's properties, along with the element's applications, and its role in biomedicine, economics, and other important subjects.

Gonick, Larry, and Craig Criddle. *The Cartoon Guide to Chemistry.* New York: Collins, 2005. Gonick has written a number of "Cartoon Guides," which explain topics such as chemistry with illustrations and brief explanations. Although the term *cartoon* may invoke a sense of childish simplicity, these guides generally offer a clear and concise introduction to all aspects of the subject.

LeCouteur, Penny, and Jay Burreson. *Napoleon's Buttons: How 17 Molecules Changed History.* New York: Tarcher, 2003. Napoleon's invasion of 1812 met with disaster in Russia's harsh winter. LeCouteur and Burreson describe how the tin buttons and fasteners of the uniforms of Napoleon's army disintegrated in the cold weather, exposing the men of this mighty army to frigid conditions and weakening their fighting ability. This and other stories show how the chemical properties of the elements and compounds have dramatically affected the course of history throughout the ages.

Strathern, Paul. *Mendeleyev's Dream: The Quest for the Elements.* New York: Thomas Dunne Books, 2001. Here is the story of the central tenet of chemistry—the periodic table of elements, the idea for which came to Russian chemist Dmitri Mendeleyev in a dream.

Web Sites

Exploratorium. Available online. URL: http://www.exploratorium.edu/. Accessed May 28, 2009. The Exploratorium, a museum of science, art, and human perception in San Francisco, has a fantastic Web site full of virtual exhibits, articles, and animations, including much of interest to chemists and chemists-to-be.

How Stuff Works. Available online. URL: http://www.howstuffworks.com/. Accessed May 28, 2009. This Web site hosts a huge number of articles on all aspects of technology and science, including chemistry.

Nobel Foundation: The Nobel Prize in Chemistry. Available online. URL: http://nobelprize.org/nobel_prizes/chemistry/. Accessed May 28, 2009. Ever since 1901, when Dutch chemist Jacobus H. van 't Hoff won the first Nobel Prize in chemistry for his work on equilibrium and the rates of chemical reactions, the Nobel Foundation has recognized and awarded important advances in chemistry. This

Web site lists the winners and provides biographical information as well as descriptions of their research.

ScienceDaily. Available online. URL: http://www.sciencedaily.com/. Accessed May 28, 2009. An excellent source for the latest research news, ScienceDaily posts hundreds of articles on all aspects of science. The articles are usually taken from press issues released by the researcher's institution or by the journal that published the research. Main categories include Fossils & Ruins, Mind & Brain, Earth & Climate, Matter & Energy, and others. Many of the chemistry articles fall in the Matter & Energy category.

WebElements: The Periodic Table on the Web. Available online. URL: http://www.webelements.com/. Accessed May 28, 2009. Comprehensive coverage of the elements on this site includes basic information, history of discovery, uses for the element, its common compounds, and much other chemical information.

INDEX